Monographs of the Physiological Society No. 29

The stress of hot environments

THE STRESS OF
HOT ENVIRONMENTS

D. McK. KERSLAKE

Consultant, Royal Air Force Institute
of Aviation Medicine

CAMBRIDGE
At the University Press 1972

K CDE

Published by the Syndics of the Cambridge University Press
Bentley House, 200 Euston Road, London NW1 2DB
American Branch: 32 East 57th Street, New York, N.Y.10022

Library of Congress Catalogue Card Number: 74-168896

ISBN: 0 521 08343 5

Printed in Great Britain
at the University Printing House, Cambridge
(Brooke Crutchley, University Printer)

CONTENTS

Contents

Contents

PREFACE

It is a matter of common experience that the air temperature alone is not an adequate indication of environmental warmth. Everyone recognizes the importance of wind, sunshine and humidity, and the notion that all these factors might be combined into a single figure indicating warmth is immediately attractive. In this monograph the problem of constructing such an index of heat stress is examined from a theoretical point of view. It is perhaps self evident that the way in which the environmental factors should be combined must depend on the properties of the subject exposed to them, but none of the heat stress indices in current use makes formal allowance for this. A quantitative examination of the importance of physiological differences between different subjects therefore seems worth while.

The first four chapters deal with the physical principles of heat exchange at the skin surface. The treatment is rather more detailed than is strictly necessary for the subsequent development, but may be of help to those who wish to penetrate the mystique which surrounds these matters.

A full scale review of the physiology of human thermoregulation would be beyond the scope of this book. Physiological theory does not yet provide an adequate basis on which an index of heat stress may be developed, and the relevant material is essentially empirical. The chapter on physiological responses presents a personal and not necessarily orthodox view of the current position, an indulgence which will be justified if it provokes controversy.

At the time of writing, the physiological world is experiencing a cathartic change to the s.i. system of units. In order to avoid using obsolete units in the text it has been necessary to modify many of the quantitative statements from the literature. It is hoped that authors who have been quoted will accept this degree of mis-representation, and that their charity will extend to any errors which may have arisen in the process of transformation.

The s.i. system offers certain alternatives, and it is too early yet to see which way the physiological cat will jump. For metabolic rates the watt seems the most appropriate unit, at least for the purposes of this book. However the time scale of adjustments to heat is such that the second is rather a small unit of time. The kilowatt hour is of dubious parentage, and the kilojoule is preferable as a unit of stored heat. Complications are inevitable if exposure times are reckoned in minutes or hours. For vapour pressures the millibar would be quite convenient, but is possibly obsolescent. The kilopascal is used here, the pascal being a name for the newton per square metre (unhappily not yet accepted internationally). A name for this unit is essential, to avoid cancelling dimensions. Thus, if the coefficient for heat exchange by evaporation is expressed in $W/m^2.kPa$ its structure is clear. To avoid the word 'Pascal' one must either write $W/m^2(kN/m^2)$, which is clumsy, or W/kN, which is incomprehensible.

Objection has been raised in the past to the expression of sweat rate in heat units (then $kcal/m^2.h$, now W/m^2). It is true that sweat is never actually measured in these units, and that unevaporated sweat does not manifest itself as removal of heat. However, the evaporative heat loss is necessarily expressed in heat units, and the important ratio of this to the sweat rate can only be dimensionless, like all the best ratios, if sweat is expressed in the same units. One could, of course, introduce the latent heat of evaporation into equations connecting these quantities, but the interest of simplicity seems better served by expressing the sweat rate as a potential rate of heat loss.

The manuscript was completed at the end of 1970, and I have not attempted to update it by reference to work published since then.

It is a pleasure to acknowledge the constant help of Mr D. F. Brebner throughout the preparation of this book. I am also grateful to the authors and publishers who have given permission for material to be reproduced here. The book could not have been written without the encouragement to fundamental research given by successive Directors General of Medical Services, Royal Air Force, and by the late Air Vice-Marshal W. K. Stewart, who directed the work of this Institute from 1947 to 1967.

D. MCK. KERSLAKE

Royal Air Force Institute of Aviation Medicine,
Farnborough
January 1972

1 HEAT EXCHANGE WITH THE ENVIRONMENT

Heat tends to pass from places where the temperature is high to places where it is lower. When it passes through solids or through fluids which are not moving the process is called conduction, when through a moving fluid (liquid or gas) it is called convection. Heat may also be exchanged across a gap between two surfaces by radiation.[1] Heat transferred by these three channels is known as sensible heat. A fourth channel of great importance in physiology is evaporation. When heat is lost by evaporation, or gained by the converse process of condensation, temperature differences are not directly involved. The heat taken up or liberated depends on the change of state of the water from liquid to vapour or vice versa, and as this is a consequence of the latent heat of vaporization, heat transferred in this way is called latent heat.

Conduction

Heat transfer by conduction follows laws which are fundamentally very simple. The driving force is temperature difference and the rate at which heat is transferred depends on this and on the thermal resistance. For the steady state, i.e. when the temperatures at all points in the system are steady, the analogy with Ohm's law is exact.

$$H = (T_1 - T_2)/R, \tag{1}$$

H is the rate of heat transfer (watts), T_1 and T_2 the temperatures at the places between which the heat is being transferred (°C) and R the thermal resistance between them (°C/W). Conductance, which is the inverse of resistance, is sometimes a more convenient term to use.

$$H = k(T_1 - T_2). \tag{2}$$

The units of conductance, k, are W/°C.

---

[1] Confusingly, in books on heat conduction the processes of convection and radiation may be lumped together and called radiation.

1. *Heat exchange with the environment*

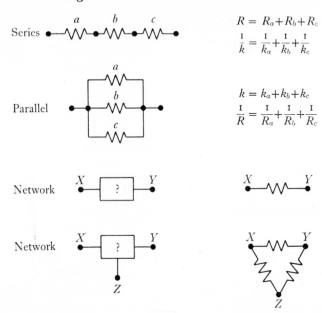

$$R = R_a + R_b + R_c$$

$$\frac{1}{k} = \frac{1}{k_a} + \frac{1}{k_b} + \frac{1}{k_c}$$

$$k = k_a + k_b + k_c$$

$$\frac{1}{R} = \frac{1}{R_a} + \frac{1}{R_b} + \frac{1}{R_c}$$

Fig. 1.1. Networks of resistances or conductances. The components a, b and c have resistances R_a, R_b, R_c; conductances k_a, k_b, k_c. ($R = 1/k$). The last two examples show unknown networks of resistances with two and three 'terminals', i.e. points at which heat may be transferred in or out of the network. Whatever the network really is, it may be represented in the form shown on the right.

Networks of thermal resistances or conductances can be combined in the same way as their electrical counterparts.

1. The equivalent resistance of a number of resistances in series is equal to the sum of the individual resistances.

2. The equivalent conductance of a number of conductances in parallel is equal to the sum of the individual conductances.

3. Any network of resistances between two points can be represented as a single resistance between the two points.

4. Any network of resistances between three points can be represented as a ring of resistances joining the three points.

Conductance and conductivity

The distinction is useful and important, but definitions of these words have an unavoidable flavour of mediaeval metaphysics which may be confusing. For example, 'The suffix "ity" denotes a fundamental, i.e. "intrinsic state or condition of" the quantity,

2

and one which is not dependent on external dimensions or conditions: The suffix "ance" denotes a specific form, condition or modification of the fundamental quantity'.

An example may make this clearer. Suppose we have two metal plates each 2 m square, separated by a slab of wood 0.1 m thick (Fig. 1.2). The temperature difference between the two plates is 10 degC, and it is observed that heat passes from one to the other at the rate of 30 W. The conductance of the slab can be found from equation (2): $k = H/(T_1 - T_2) = 3.0$ W/°C. In physiology heat transfer is often expressed per unit surface area. In this example,

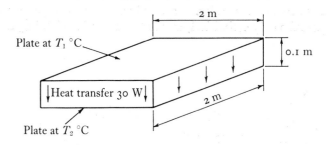

Fig. 1.2. Two metal plates, each 2 m square are separated by a slab of wood 0.1 m thick. The temperatures of the two plates are T_1 and T_2, and the temperature difference is 10 degC. Heat flows through the wooden slab at the rate of 30 W.

since each plate has an area of 4 m², the heat passes through this area of wood, and the conductance per unit area is 0.75 W/m². °C. Although the area term has been introduced this is still a conductance, because the dimension of thickness of the slab is still present. To remove this, and to arrive at the conductivity of the wood, the thickness must be standardized to the dimension of length in which we are working, in this case the metre. Clearly the heat transfer through a slab 1 m thick would be a tenth of that through the 0.1 m slab in the example. (Thinking in terms of resistance we have a pile of ten slabs in series.) The conductance of a slab 1 m thick would thus be 0.075 W/m². °C, and the conductivity of the wood is 0.075 W.m/m². °C. The change in the units is due to the removal of the qualification about the thickness of the slab. One of the length dimensions can be cancelled so that conductivity is expressed as W/m. °C.

3

1. *Heat exchange with the environment*

Steady state conduction in solids

In the above example of heat conduction through a slab it was tacitly assumed that there was no heat exchange through the edges of the slab. At all points in the slab the heat flowed in the same direction. In a formal treatise on heat conduction this would be the case of a slab of infinite area. Conduction through a rectangular block in which heat transfer occurs at all six faces is a far more complicated problem beyond the scope of this book (see, for example, Carslaw & Jaeger, 1959).

A case of particular interest in physiology is that of conduction radially through a cylinder. Most body segments can be regarded as approximately cylindrical, as can the overlying clothing, and it

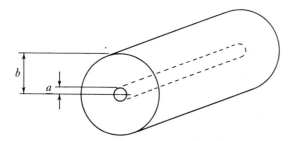

Fig. 1.3. A cylindrical body segment of radius *a* covered with clothing of outer radius *b*.

is useful to be able to calculate the thermal resistance of a cylindrical layer of clothing. Consider a body segment of radius *a* covered by a layer of clothing of outer radius *b* (Fig. 1.3). For unit length of the segment the surface area of the skin is $2\pi a$, and if heat is lost at the rate *H* watts per unit length, the flux of heat at the inner surface of the clothing is $H/2\pi a$ W/m². The area of the outer surface of the clothing is greater, and since the total rate of heat transfer is the same the heat flux here must be less. The temperature gradient in the outer layers must also be less because the flux is less. At any point in the clothing distant *x* from the centre, the temperature gradient is given by

$$dT/dx = -H/K.2\pi x,$$

where *K* is the thermal conductivity of the clothing material. The temperature difference between the inner and outer surfaces of the clothing may be found by integrating between the limits $x = a$ and $x = b$.

$$T_a - T_b = \frac{H}{2\pi K}.\ln(b/a).$$

The thermal resistance of the clothing, $(T_a - T_b)/H$, is given by

$$R = \frac{1}{2\pi K}.\ln(b/a).$$

At first sight this is rather an odd result. The thermal resistance depends on the ratio b/a whereas the thickness of the clothing layer, $(b-a)$, does not appear. Furthermore as the radius, a, increases and b/a for a given clothing thickness approaches unity, the thermal resistance approaches zero. This is in fact quite reasonable because if b/a were unity the surface areas of both the skin and the clothing would be infinite and the fixed quantity of heat, H, would have no difficulty in escaping. The surface area and clothing thickness are related in the ratio b/a.

Fishenden and Saunders (1950, p. 42) point out that the thermal resistance of cylindrical insulation can usually be found with adequate accuracy by a simpler method. The clothing is considered as a flat slab of area equal to the arithmetic mean of the inner and outer surface areas, $\pi(a+b)$, and of thickness $(b-a)$.

$$R = (b-a)/K\pi(a+b).$$

The error is only about 4 per cent when b/a is 2, and the thickness of clothing is rarely greater than the radius of the underlying body segment.

Conduction in non-steady states

The conduction of heat in the steady state is a special case of the more general problem of the conduction of heat in solids. The simple analogy with electrical resistance only applies in the steady state, when the heat content of the material is not changing. Conduction in the non-steady state will not be considered in detail here, since it is a subject of considerable complexity with few direct applications in physiology. The reasons for the complexity will be illustrated by examining the basic equation of heat conduction.

1. *Heat exchange with the environment*

Consider a slab of material through which heat is flowing (Fig. 1.4). A temperature gradient, dT/dx, will exist at any point, and the rate of heat transfer per unit cross sectional area at that point will be $K.dT/dx$, where K is the thermal conductivity of the material. In the case of the region bounded by x_1, x_2, the rate of entry of heat is K times the temperature gradient at x_1, and the

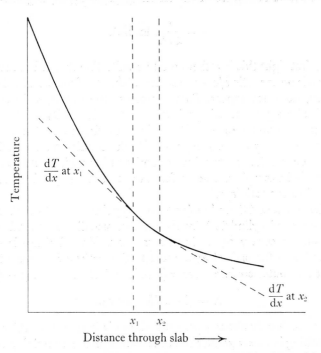

Fig. 1.4. Instantaneous distribution of temperature in a slab of material which is not in the steady state. The temperature gradients at the points x_1 and x_2 are indicated by the dotted lines, which are tangents to the temperature curve at these points. Heat enters and leaves the portion of the slab bounded by x_1 and x_2 at rates proportional to these gradients, and since the rates are unequal the temperature changes with time.

rate of exit is the corresponding expression at x_2. The difference between these is the rate of acquisition of heat by this part of the slab and can be expressed for the case of a very thin slice

$$dQ/dt = K.d^2T/dx^2.$$

Because heat is stored in the material, the temperature will rise and the rate of rise will depend on the rate of heat storage and the heat capacity. The heat capacity per unit volume is $\rho.c$, where ρ is the density and c the heat capacity per unit mass. Thus the rate of change of temperature is given by

$$dT/dt = \frac{K}{\rho.c}.d^2T/dx^2.$$

This is the fundamental equation for the linear flow of heat. The problem is to solve it so as to express temperature, time, distance and their combinations in terms of one another. Various mathematical approaches can be used, but the solutions, even for very simple cases, are elaborate and difficult to manipulate. Heat transfer in solids can be regarded as a process of diffusion, and the equations describing it are similar to those relating to other diffusion processes. (Ingersoll, Zobel & Ingersoll, 1954; Crank, 1956; Carslaw & Jaeger, 1959.)

Convection

The exchange of heat between a surface and a fluid (liquid or gas) in contact with it takes place by conduction through the fluid, but is complicated by movement of the fluid. Under conditions of interest in physiology the fluid near the skin is exchanged rapidly, and so far as the fluid itself is concerned the process can be regarded as in the steady state. (This is so even though the surface may not be in the steady state.) The expression for heat exchange by convection is similar to equation (2) for conduction. Writing C for the rate of heat exchange per unit area,

$$C = h_c(T_1 - T_2). \tag{3}$$

Here h_c is the coefficient for heat transfer by convection (convection coefficient). Its units are $W/m^2.°C$, and it behaves just like any other thermal conductance. Its value depends on the nature of the fluid and how it is flowing (Chapter 2).

Radiation

Heat can be exchanged between surfaces by radiation. The intervening air is not involved, and heat can be radiated across empty space. The process is considered in more detail in Chapter 3. For the present it is sufficient to indicate that although the process depends on the difference between the fourth powers of the abso-

1. *Heat exchange with the environment*

lute temperatures of the two surfaces, an approximation is often permissible whereby the rate of heat transfer is related to the temperature difference (first power). When this approximation is used, heat transfer by radiation can be represented in the same form as that for conduction and convection.

$$R = h_r(T_1 - T_2). \tag{4}$$

Here R is the rate of heat transfer per unit area, h_r is the coefficient for heat exchange by radiation (W/m^2.°C) and T_1 and T_2 the temperatures of the two surfaces. The coefficient h_r depends on the natures of the two surfaces, their temperatures and the geometrical relation between them.

When this first power approximation is used for radiant heat exchange, the three processes of sensible heat exchange, conduction, convection and radiation, follow mathematically similar laws, equations (2), (3) and (4).

The effect of distance on heat radiation is sometimes misunderstood. For example, a gardening book, referring to the radiation of heat from walls, says that the wall 'radiates its heat in a ratio proportionate to the square of the distance; so that if an object placed a foot from the wall receives 1 deg. of heat from it, at 1 inch it will receive heat equal to 144 degrees'. The confusion between temperature and heat is incidental; the most important point is the application of the inverse square law to this situation. It is true that the radiant flux from a point source diminishes as the square of the distance. Heat radiation, however, usually comes from extended surfaces, and diminishes as the solid angle subtended by the radiating surface. If the wall is at all large the solid angle will scarcely change over the distances quoted. It is warmer near the wall mainly because the air movement is lower there, and heat received from the wall is more effective in warming the plant (see Operative temperature, p. 66).

Evaporation

Although the removal of heat by evaporation depends on the change of state of water from liquid to vapour, the rate at which the process proceeds is determined by the rate at which the vapour diffuses away from the surface. Conduction and convection can both be regarded as processes of diffusion of heat from regions of high concentration (temperature) to regions of lower concentration, and evaporation follows laws very similar to those governing convection. Physiologists conventionally express evaporation in terms of the latent heat taken up and the vapour pressure difference which constitutes the driving force for diffusion.

$$E = h_e(p_1 - p_2). \tag{5}$$

E is the rate of heat loss by evaporation per unit area, h_e the evaporation coefficient (W/m^2.kPa), analogous with h_c for convection, and p_1 and p_2 the partial pressures of water vapour at the skin surface and in the ambient air.

An important concept is that of saturation of air with water vapour. Imagine a closed jar of moist air in which there is a pool of water at the same temperature as the air. Molecules in the water are in motion and some of those at the surface may move fast enough to escape from it and enter the air as water vapour. At the same time molecules of water in the atmosphere, which are also in motion, will sometimes return to the water surface and re-enter the liquid phase.

If the concentration of water molecules in the air is high enough the rates of evaporation and condensation will be equal. This concentration, when expressed as a partial pressure, is the saturated water vapour pressure at the temperature of the water surface. Air at this temperature, containing this concentration of water vapour, is said to be saturated. The warmer the water, the faster its molecules move and the greater the rate at which they escape. The saturated water vapour pressure therefore increases with temperature. The curved line in Fig. 1.5 shows the relation between temperature and saturated vapour pressure. Values are tabulated in Appendix 1.

If the water vapour pressure in the air is below saturation, a pool of water at air temperature would evaporate into the air, and in a closed system equilibrium would eventually be reached at the saturated vapour pressure. If the water vapour pressure in the air exceeded saturation (as, for example, if damp air were suddenly cooled) the excess of water vapour would condense as dew on surfaces or as mist in the air. (There are some niceties about super-saturation and the vapour pressure of small droplets which complicate the picture in practice, but which do not affect the fundamental principles.)

9

1. *Heat exchange with the environment*

It is noteworthy that evaporation can occur into saturated air if the evaporating surface is above air temperature.

Physiologists usually express humidity as the partial pressure of water vapour in the air, but other terms may be used in other contexts and are here listed for the purposes of comparison and conversion. (Penman, 1955; Flink, 1960.)

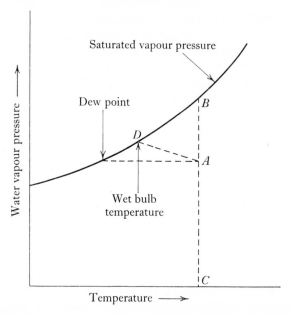

Fig. 1.5. Various ways of expressing the humidity of the environment indicated by the temperature and water vapour presure of the point, *A*. The curved line shows saturated water vapour pressure. Dew point is the temperature at which the prevailing water vapour pressure would be equal to the saturated water vapour pressure. Relative humidity is the ratio *AC/BC*. Saturation deficit is the vapour pressure difference *AB*. Wet bulb temperature is the temperature reached if water is evaporated into the air adiabatically until saturation is reached.

Dew point is the temperature at which dew would begin to form if the air were slowly cooled. It is therefore the temperature at which the saturated water vapour pressure is equal to the prevailing partial pressure of water vapour in the air and is an alternative way of expressing the partial pressure. In Fig. 1.5 the dew point of the environment *A* is at the junction of the horizontal line through *A* (constant vapour pressure) and the curve for saturated vapour pressure. Dew will form on a surface cooled below the dew point, since the air in immediate contact with the surface will be cooled to this temperature. The dew point can be directly measured by instruments working on this principle.

Relative humidity is the ratio of the prevailing partial pressure of water vapour to the saturated water vapour pressure at the prevailing temperature, *AC/BC* in Fig. 1.5. The relative humidity is often expressed as a percentage. Some instruments, including hair hygrometers, give a direct approximate reading of relative humidity.

Saturation deficit is the difference between the saturated vapour pressure at the prevailing temperature and the actual vapour pressure, *AB*, in Fig. 1.5. It is useful in considering evaporation from a wet surface at air temperature (e.g. from the surfaces of leaves), but has no direct relation to evaporation from surfaces at any other temperature.

Absolute humidity is a measure of the mass concentration, or density of water vapour. The units are mass per unit volume of moist air, kg/m^3. This quantity should properly be used in considering evaporation from wet surfaces, although vapour pressure (p_a) will frequently serve. The two are connected by the gas laws.

$$\text{Absolute humidity} = 2.17\, p_a/T \text{ kg/m}^3,$$

if p_a is in kPa and T in °K.

Specific humidity is the mass of water vapour per unit mass of moist air. It is not in common use, since the humidity ratio (see below) is generally more convenient. The term has unfortunately sometimes been used to mean humidity ratio.

Humidity ratio (or mixing ratio) is the mass of water vapour per unit mass of dry air. It is widely used by air conditioning engineers because the mass flow of dry air in an air processing plant (apart from by-passes) is the same at all points, irrespective of temperature, pressure and humidity. Unless water vapour is added or removed, the humidity ratio is constant.

Degree of saturation (*or percentage saturation*), somewhat akin to relative humidity, though not identical with it, is the ratio of the actual humidity ratio to the saturated humidity ratio at the prevailing temperature and pressure.

Wet bulb temperature. Air can be cooled by allowing water to evaporate into it. If no other source of heat is supplied, the heat required to evaporate the water will be supplied by the air, the temperature of which will fall. Eventually the air will become saturated both because water vapour is being added and the temperature is falling. If we arrange that the liquid water is supplied at the final saturation temperature, this temperature will be the true, or thermodynamic wet bulb temperature of the air.

The line AD in Fig. 1.5 shows how the temperature and vapour pressure would change during this wetting and cooling process. Any environment on this line would have the same wet bulb temperature as environment A.

A real wet bulb thermometer would eventually reach the true wet bulb temperature if heat exchanges by radiation and stem conduction could be prevented. If the bulb is screened from radiation and a moderate air velocity is maintained across it these effects are minimized and the reading approximates to the true wet bulb temperature. The air velocity must not be too high, or kinetic heating may exert a significant effect. The wet bulb temperature depends on air temperature (also called dry bulb temperature) as well as humidity, and although the wet bulb thermometer is a convenient instrument for measuring humidity, it does not provide a direct indication of the amount of water vapour in the air.

Appendices 1 and 2 enable the water vapour pressure to be found from the wet and dry bulb temperatures. The relations between air temperature, water vapour pressure and wet bulb temperature are summarized in the psychrometric chart, Appendix 3.

Heat balance

Rates of accumulation and loss of heat can be directly added. The general equation for heat balance, which does not necessarily imply the steady state, has been represented in various essentially similar forms, but authors have differed in the matter of algebraic signs used for gains and losses. Since heat exchange by evaporation is normally a loss from the body, it is convenient to express it and the sensible heat exchanges as positive if heat is lost from the body. It also seems appropriate to put metabolic heat production on one side of the equation and to put on the other side all the places to which it may go, including retention in the body (storage), all being expressed as positive so that their sum will be equal to the metabolic heat production. In this form,

$$M - W = E + R + C + K + S, \tag{6}$$

where all the symbols represent rates of production, loss or storage, and all are expressed per unit body surface area. M is the rate of metabolic energy production and W the rate of external working. The difference, $M - W$, is the rate of production of heat in the body. S is the rate of heat storage in the tissues, and E, R, C and K the rates of heat loss by evaporation, radiation, convection and conduction respectively. A gain of heat by any of these channels is a negative loss of heat. The use of the symbol ' $\pm$ ' between the terms contributes nothing and is a potential source of confusion. Heat loss from the respiratory tract is sometimes put in as a separate term. In equation (6) it contributes to the terms E and C.

The total surface area of the body is difficult to measure directly, and it is usual to estimate it from the formula of DuBois & DuBois (1915)

$$A_D = 0.00718 \ W^{0.425}.H^{0.725}, \tag{7}$$

where A_D is the area in m², W the weight in kg and H the height in cm. Values are given in Appendix 6. The area so estimated is known as the DuBois area.

Subsequent work has resulted in improved formulae which give somewhat better estimates than equation (7). Inadequacies of this equation were noted by Boyd (1935), and the general problem of predicting surface area from height and weight is discussed by Sendroy & Cecchini (1954), who conclude that the DuBois formula is satisfactory over the greater part of the range. An improved nomogram for determining surface area was published later (Sendroy & Collison, 1960). Van Graan (1969) concludes that equation (7) underestimates the surface area by about 7 per cent.

The inadequacies of the DuBois formula are of little practical consequence in thermal physiology provided that this formula is used throughout. Determinations of heat exchange coefficients and metabolic rates are based on measurements for the whole body. These are divided by the DuBois area in order to remove most of the effects of differences of subject size, and provided that this area is proportional to the true surface area any systematic error will cancel out when heat exchange equations are applied. As will be shown in Chapter 2, convection and evaporation coefficients expressed per unit area are not the same for large and small objects of the same shape, so that differences in size cannot be fully allowed for merely by adjusting for surface area. In very accurate work it would be necessary to adjust the coefficients for subjects of different sizes. Each size of subject would then be a class of its own, and again it would not matter what formula was used for surface area.

The steady state

Equation (6) is a general statement which is true at all times. The steady state is not necessarily identified by the case $S = 0$, although this is a necessary condition. Thus if an object were plunged for a few seconds into hot water, removed and then completely insulated, the net rate of heat storage after insulation would be zero, but the temperatures in different parts of the object would continue to change until they were all equal. In the same way one can imagine conditions in which although a man's total heat loss was equal to his rate of heat production the temperatures in different parts of his body might be changing, as might the partition of heat loss between evaporation and the sensible heat transfer channels. The steady state is properly identified with unchanging temperatures, and, although rarely encountered in practice, is a useful abstraction to which real cases can be related.

2 CONVECTION AND EVAPORATION

The way in which a fluid flows past an object of complex shape defies analysis by the methods of classical physics. The approach which has been developed to deal with this problem does not attempt to examine the movement of the fluid in detail, but seeks to establish relations which will describe situations in which fluid flow and the processes of heat exchange which depend on it will be similar. The theory of similarity is characterized by the arrangement of physical quantities into dimensionless groups (ratios) which are identified by the names of their progenitors.

The essence of this approach is the description of systems which are similar, and its power lies in its ability to predict circumstances which will be similar if the size of the body or the nature of the fluid is changed. Its weakness, from the standpoint of physiology, is that it cannot compare cases in which the shapes of the objects are different (clearly the cases cannot be regarded as similar unless the shapes are so closely related that they can be described simply, as by aspect ratio) nor can it predict the magnitude of a process such as heat exchange unless this is already known empirically for a mathematically similar case. These limitations are formidable in the present context, since human subjects vary little in size (and if different in size are usually different in shape), and in most situations the surrounding fluid is air at normal atmospheric pressure, the properties of which do not vary much. The brief survey of the physics of heat transfer in fluids which follows is neither rigorous nor exhaustive. A number of corners have been cut, and with them much of the elegance of the formal treatment, to which the books by Fishenden & Saunders (1950), McAdams (1942) and Jakob (1949) provide an introduction.

The pattern of the flow of a fluid in relation to a solid body, whether a tube containing flowing water or a man standing in a wind, is determined by the shape of the solid body, the velocity of fluid flow and the physical properties of the fluid. The flow of

2. *Convection and evaporation*

air around a man clearly will not have the same pattern as that round a horse for example, but similar patterns may exist at appropriate wind speeds round large and small men or large and small horses. Shape is a property which can rarely be expressed quantitatively, but a single dimension of length, L, (any one we care to choose) will describe the size of an object of given shape.

TABLE 2.1. *Properties of air, 0–80 °C, sea level pressure.*

Temperature τ °C	Dynamic viscosity μ N.s/m²	Density ρ kg/m³	Kinematic viscosity ν m²/s	Thermal conductivity K W/m.°C
0	$1.71.10^{-5}$	1.29	$1.32.10^{-5}$	$2.41.10^{-2}$
10	1.76	1.25	1.41	2.49
20	1.81	1.21	1.50	2.56
30	1.86	1.16	1.59	2.64
40	1.90	1.13	1.69	2.71
50	1.95	1.09	1.78	2.79
60	2.00	1.06	1.88	2.87
70	2.04	1.03	1.98	2.94
80	2.09	1.00	2.09	3.02

Heat capacity per unit mass 1000 J/kg.°C
Prandtl number ≃ 0.7

If the flow patterns round two objects of the same shape but different size are similar, the ratios of the velocities of the fluid at corresponding points in the two cases will be the same. Thus, just as size can be defined by any linear measurement when shape is the same, velocity can be defined by the velocity at any appropriately defined point when the flow pattern is the same. It is usually convenient to choose the velocity of the undisturbed stream well away from the object. The pattern of fluid flow is determined by inertial and viscous forces and if the pattern of flow is to be similar in two cases the ratios of these forces at corresponding points must be the same. The Reynolds number is proportional to the ratio of the inertial forces (ρV^2) to the viscous forces (proportional to $\mu . V/L$), where ρ is the density of the fluid, μ its dynamic viscosity, V its velocity and L the characteristic dimension describing the size of the object. It can be expressed as $V.L/\nu$, where ν is the kinematic viscosity of the fluid, μ/ρ. Values of μ, ρ and ν are shown in Table 2.1 for air at normal atmospheric pressure over the tempeature range 0–80 °C. s.i. units have been used in the table, and in calculating the Reynolds number with these figures L should be in metres and V in m/s. Any consistent system of units will provide the same Reynolds number, because the quantities comprising it are such that the number itself is dimensionless. Its magnitude is arbitrary, since L can be any characteristic length. L could be the height of a man or the length of his nose, since in men of the same shape these quantities are in the same ratio. If the Reynolds number is the same for two men of different size, the pattern of fluid flow around them is similar, and this is true (within limits) for any fluid.

Some confusion may arise from statements involving absolute values of Reynolds number. For example, it is true that the flow in a long tube becomes turbulent if the Reynolds number exceeds about 3000, provided that the diameter of the tube is taken as the characteristic dimension. If the radius were taken instead the figure would be halved, but it is conventional to use the diameter. While academically objectionable it is even possible to assign an equivalent diameter to a rectangular duct, when the behaviour is much the same as for a circular duct of equivalent size. However in general when objects are of dissimilar shape no comparison between them based on Reynolds number is reliable or justifiable.

Convection

Forced convection

The value of the convection coefficient, h_c, in the equation, $C = h_c(T_s - T_a)$, will clearly depend on the flow pattern and the thermal properties of the fluid. If the exchange of heat alters the density of the fluid the flow pattern may be affected, and when this process dominates the flow the condition is called natural convection. The case of forced convection, in which the effect of changes in density on the flow can be neglected (i.e. at high imposed fluid velocities) is simpler. Here the Reynolds number will specify the flow pattern, and two other dimensionless numbers, the Nusselt and Prandtl numbers, will take care of the heat exchange and the thermal properties of the fluid.

The Nusselt number is $h_c.L/K$, where h_c is the coefficient for heat exchange by convection (W/m². °C), L is the characteristic dimension of size as before, and K is the thermal conductivity of

2. *Convection and evaporation*

the fluid. The Nusselt number is proportional to the ratio of the actual coefficient for heat transfer by convection, h_c, to that by conduction in the same fluid at rest (proportional to K/L). The coefficient of expansion of the fluid is assumed to be without effect in forced convection, and the remaining thermal property, specific heat, is contained in the Prandtl number, $\mu.c/K$, where μ is the dynamic viscosity, c the heat capacity per unit mass (at constant pressure) and K the thermal conductivity. The Prandtl number is the ratio of the kinematic viscosity, μ/ρ, to the thermal diffusivity, $K/\rho c$. Kinematic viscosity determines the way in which momentum is transferred across a velocity gradient and thermal diffusivity determines the way in which heat is transferred across a temperature gradient. The Prandtl number connects momentum transfer (in the Reynolds number) with heat transfer (in the Nusselt number). The relation between heat transfer and fluid flow can be expressed in the functional notation,

$$N_{\mathrm{Nu}} = f(N_{\mathrm{Re}}, N_{\mathrm{Pr}}). \tag{1}$$

The Prandtl number contains only physical properties of the fluid and is uninfluenced by flow pattern, velocity, etc. If the nature of fluid is fixed, the Prandtl number can be ignored in examining the relation between fluid velocity and heat transfer. The pattern of fluid flow is identified by the Reynolds number, so that for two bodies of the same shape in air, if the Reynolds number is the same the Nusselt number must also be the same.

This statement is of rather limited application because although the two dimensionless numbers determine one another the relation between them is not defined. It is clear that this must be so, since otherwise, neither number containing a statement of the shape of the body in question, shape would be without effect on heat transfer. For any given shape the numbers are related, but this relation must be discovered by experiment. There is no reason to expect it to have any simple mathematical form, and any formulation will be strictly empirical. Because the Reynolds and Nusselt numbers are dimensionless there is no objection to taking logarithms, one of the traditional ways of straightening out empirical data. It is therefore convenient to express their relation by a power law of the form

$$N_{\mathrm{Nu}} = B.N_{\mathrm{Re}}^n. \tag{2}$$

An example of a relation between Nusselt and Reynolds numbers is illustrated in Table 2.2, which shows values for B and n in equation (2) determined by Hilpert (1933) for cylinders transverse to a wind. The plot of $\log N_{\mathrm{Nu}}$ against $\log N_{\mathrm{Re}}$ is interpretable as the series of straight lines defined in Table 2.2. It is impressive that the same curve was well fitted by observations on cylinders from 0.02 to 150 mm diameter. When the Prandtl number is included (raised to the power 0.3 and multiplied on the right-hand side of equation (2)), observations on heat transfer in liquids fit closely those in gases (Fishenden & Saunders, 1950, p. 129). The Reynolds and Nusselt numbers both contain the characteristic dimension, L, which is arbitrarily chosen, and which would only cancel out in the case $n = 1.0$, which does not occur. The values of B in Table 2.2 therefore depend on the dimension which has been chosen for L (in this case it was diameter), and where there is no generally accepted convention it is necessary to discover what choice was made before information presented in the form of equation (2) can be applied.

TABLE 2.2. *Constants of equation (2) for forced convection perpendicular to cylinders. Over the range* $N_{\mathrm{Re}} = 1000–100000$, $B = 0.24$, $n = 0.60$ *is almost correct (Fishenden & Saunders, 1950, p. 130).*

N_{Re}	B	n
1–4	0.891	0.330
4–40	0.821	0.385
40–4000	0.615	0.466
4000–40000	0.174	0.618
40000–250000	0.024	0.805

Natural convection

If a warm object is placed in still air, convection currents are generated as the air near the surface is warmed, expands and therefore rises. It is easy to see that in these circumstances of natural convection the movement of the air will depend on the temperature difference between the surface and the air, the coefficient of expansion of the air and the gravitational field, as well as on the properties which are involved in forced convection. No velocity term is necessary in equations describing natural convection because the velocities are generated by the heat transfer process and are not imposed from outside. The additional quantities are combined in the Grashof number, $g.L^3.\beta.\theta/\nu^2$, where g is the acceleration due to gravity, β the coefficient of expansion of the air (equal

2. Convection and evaporation

to $1/T$, where T is the absolute temperature (Fishenden & Saunders, 1950, p. 73)) and θ the temperature difference between the surface and the air. As in the case of forced convection (equation (1)) the Prandtl number is present in the full equation

$$N_{Nu} = f(N_{Gr}, N_{Pr}) \qquad (3)$$

but may be omitted if only air environments are considered.

$$N_{Nu} = B' . N_{Gr}^{n'}. \qquad (4)$$

The empirical constants, B' and n' in this equation are, of course, unrelated to those in equation (2), and must be determined by experiment for the shape in question. For many shapes at large Grashof numbers (large convection velocities) n' in equation (4) is about $\frac{1}{3}$. Since the Grashof number contains L^3, its $\frac{1}{3}$ power varies directly as L. The Nusselt number is also directly proportional to L, so that the heat transfer coefficient, h_c, is independent of size. At smaller values of Grashof number the exponent is smaller, and smaller objects have larger heat transfer coefficients at the same Grashof number. The change occurs when the convection currents become turbulent. For cylindrical bodies at Grashof numbers below 10^8, the expression

$$N_{Nu} = 0.47 \, (N_{Gr}. N_{Pr})^{0.25}$$

is adequate.

Natural and forced convection

For human subjects at wind speeds above about 0.2 m/s, forced convection predominates. Below this speed natural convection may contribute significantly to heat exchange, its magnitude depending on the temperature difference between skin and air. Lewis *et al.* (1969) have demonstrated natural convection in human subjects by Schlieren photography and have measured the air velocities in the boundary layer. In the transition zone between natural and forced convection it is usual to calculate the heat transfer coefficient for each process and to use the larger of the two values (Fanger, 1967). More elaborate methods for combining natural and forced convection are discussed by Sibbons (1970).

Evaporation

Heat transfer by convection is essentially a process of diffusion of heat through the boundary air layer near the skin surface. Evaporation requires an initial change of state from liquid to vapour at the skin surface and the subsequent diffusion of the vapour across the boundary air layer into the ambient air. The two processes thus have much in common, and provided that the volume of water vapour is not great enough to disturb the pattern of air flow they follow very similar laws. The driving force for diffusion is the concentration gradient, and the quantity transferred is mass. The rate of mass transfer per unit area is described by

$$\dot{m} = h_D(C_s - C_a), \tag{5}$$

where C_s and C_a are the water vapour concentrations (mass per unit volume) at the skin surface and in the ambient air and h_D is the mass transfer coefficient, somewhat analogous to h_c, the convection coefficient, but with dimensions m/s, since it is multiplied by concentration, kg/m^3, to yield $\dot{m}$ (kg/m^2.s).

The mass transfer coefficient, h_D, is contained in the Sherwood number, $h_D.L/D$, analogous to the Nusselt number in convective heat transfer, $h_c.L/K$. D is the mass diffusivity, and one might expect by analogy to find thermal diffusivity $(K/\rho.c)$ rather than conductivity, K, in the Nusselt number. The reason for the difference lies in the dimensional structure of the driving forces in the two cases, since both numbers must be dimensionless. As with convection, the Reynolds number describes the flow pattern, but the Schmidt number, ν/D, replaces the Prandtl number, ν/α (α is the thermal diffusivity).

$$N_{Sh} = f(N_{Re}, N_{Sc}). \tag{6}$$

It so happens that D for water vapour diffusing through air is nearly equal to α, so that the Prandtl and Schmidt numbers are nearly equal for this case. The analogy between water vapour transfer and sensible heat transfer in air is therefore a particularly close one, the Sherwood and Nusselt numbers are nearly equal, and the coefficients h_D and h_c are related by the equation

$$h_D \simeq h_c/\rho.c. \tag{7}$$

This relation is sometimes known as Lewis's rule (Lewis, 1922), and the ratio $h_c/h_D.\rho.c$, approximately unity in the present case, is called the Lewis number.

It follows from the similarity between convective heat transfer and evaporation (whatever the value of the Lewis number) that

2. Convection and evaporation

both processes depend in the same way on the Reynolds number, and that for any object the coefficients will bear the same ratio to one another at all wind speeds.

Effects of pressure and temperature
Atmospheric pressure

The Prandtl number is independent of temperature and pressure over a wide range, so forced convective heat transfer at different atmospheric pressures can be adequately represented by equation (2). Expressing this in terms of the basic quantities involved,

$$\left. \begin{aligned} h_c.L/K &= B(V.L.\rho/\mu)^n, \\ h_c &= B.K.V^n.L^{(n-1)}.\rho^n.\mu^{-n}. \end{aligned} \right\} \tag{8}$$

For air over the range of atmospheric pressures tolerable by Man, μ and K may be regarded as independent of pressure, but at constant temperature ρ is directly proportional to pressure. If V is fixed and the pressure changes, h_c varies directly as ρ^n, i.e. it increases with increasing pressure. An alternative way of looking at this is to say that h_c is a function of the mass velocity, ρV.

The effect of pressure on evaporation can be examined in the same way. Like the Prandtl number, the Schmidt number, ν/D, is independent of pressure. In terms of the basic quantities the equation, $N_{Sh} = B.N_{Re}^n$, gives

$$h_D = B.D.V^n.L^{(n-1)}.\rho^n.\mu^{-n}, \tag{9}$$

D is inversely proportional to ρ, so h_D varies as $\rho^{(n-1)}$. If n is less than unity, h_D will diminish with increasing density (increasing pressure). h_c varies as ρ^n, and the ratio h_c/h_D must be directly proportional to density (cf. equation (7)). The opposite effects of pressure on convection and evaporation make sense, since extra air molecules might be expected to promote sensible heat exchange while hindering the diffusion of water molecules.

Temperature

The Prandtl number for air is independent of temperature, but at constant pressure both ν, which occurs in the Reynolds number, and K, which occurs in the Nusselt number, change with temperature (Table 2.1). K varies approximately as the absolute temperature and ν as the square of the absolute temperature. If ν is substituted for μ/ρ in equation (8) for forced convection,

$$h_c = B.K.V^n L^{(n-1)}.\nu^{-n}. \tag{10}$$

Estimates of the value of n in this equation for human subjects at low and moderate wind speeds lie between 0.67 and 0.50 (Table 2.5), the latter value being commonly assumed. For this value, equation (10) shows that h_c varies as $K/\sqrt{v}$, which is constant. Thus in the case of human subjects the coefficient for heat transfer by convection at constant pressure is almost independent of the general temperature level. In many engineering cases the value of n is about 0.8, and temperature exerts a significant effect.

Equation (9) for evaporation, can be expressed

$$h_D = B.D.V^n.L^{(n-1)}.v^{-n}. \tag{11}$$

Both D and v vary approximately as the square of the absolute temperature, so that for the case $n = 0.5$, the mass transfer coefficient, h_D, varies directly as the absolute temperature. As will will be shown in the next section, when evaporation is expressed in terms of vapour pressure rather than vapour concentration, the dependence on temperature disappears.

Evaporation and vapour pressure

When evaporation is treated strictly as a diffusion process, the driving force is the concentration gradient and the rate of evaporation is expressed as the rate of transfer of mass (equation (5)). In physiology it is usual to express the driving force as water vapour pressure and the rate of evaporation as the heat to which it is equivalent.

$$E = h_e(p_s - p_a). \tag{12}$$

The evaporation coefficient, h_e, replaces the mass transfer coefficient, h_D, in equation (5), and its units are $W/m^2.kPa$. The vapour pressure, p, is related to the vapour concentration, C, in equation (5), by the gas laws. The volume of a given mass of gas depends on the pressure and the absolute temperature. C is the mass per unit volume and equals p/RT. It was shown above that for the case of a human subject h_D is roughly proportional to the absolute temperature. Substituting $k.T$ for h_D and p/RT for C in equation (5),

$$\left. \begin{aligned} \dot{m} &= k.T(p_s/RT - p_a/RT), \\ \dot{m} &= (k/R)(p_s - p_a). \end{aligned} \right\} \tag{13}$$

Thus when vapour pressure is used instead of vapour concentration, the mass rate of evaporation is independent of temperature.

2. *Convection and evaporation*

(These formulations for evaporation apply strictly to the case in which there is no temperature difference between the skin and the air, but moderate temperature differences do not exert an important effect.)

The evaporative heat loss, E, is equal to $\dot{m}.\lambda$, where λ is the latent heat of vaporization, the heat required to convert water into saturated vapour at the same temperature. λ varies a little with temperature, but the process of evaporation takes place at skin temperature, and over the range 30–38 °C the change in λ is only about 1 per cent. For practical purposes, therefore, h_e in equation (12) may be regarded as independent of temperature. Equation (12) presupposes that the skin is completely covered with a film of water, and it is only in this case that E is equal to $\dot{m}.\lambda$. The heat taken up by evaporation in other circumstances is examined below.

The total heat of evaporation

The latent heat of vaporization is the heat required to convert liquid water into saturated vapour at the same temperature. This involves a change of volume, and work must be done against the external pressure (saturated vapour pressure, not total atmospheric pressure). The heat equivalent of this work, known as the external work, is included in the latent heat of vaporization. This does not complete the process, however, since after diffusing away from the surface, the water vapour finishes up at ambient partial pressure and ambient temperature, additional heat exchanges being required to bring this about. These have been treated by Hardy (1949) and Taylor (1954). In a closed calorimeter the total heat taken up by evaporating water is the latent heat of vaporization plus the heat involved in the subsequent processes (Murlin & Burton, 1935). The heat required to change the phase of the water is removed directly from the skin, but some of the rest may be derived from the air in the calorimeter. The system is closed, and the total heat transfer will work out the same whatever assumption is made about the source of heat for the secondary processes, because cooling of the calorimeter air will affect the calculated sensible heat exchange.

In an open system, on the other hand, a problem arises, since if some of the heat is taken up from air which is swept away, not all the heat involved in the evaporation will be abstracted from the subject. Clearly the latent heat of vaporization is derived from the

24

skin, since the air in immediate contact with the skin is at skin temperature. Some indication of the importance of the secondary processes is given by the behaviour of the wet bulb thermometer. The relation between the wet bulb depression (dry bulb temperature minus wet bulb temperature) and the vapour pressure difference (saturated vapour pressure at wet bulb temperature minus

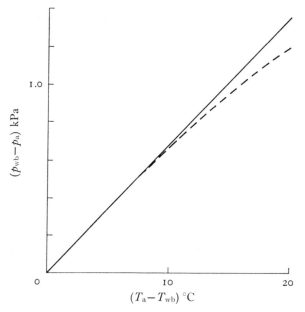

Fig. 2.1. The relation between wet bulb depression $(T_a - T_{wb})$ and vapour pressure difference $(p_{wb} - p_a)$ for the wet bulb thermometer. The continuous line indicates the actual relation, the dashed line the relation which would be expected if the heat taken from up the wet bulb were equal to the total heat of evaporation.

ambient water vapour pressure) is shown by the straight line in Fig. 2.1. The actual equation is

$$(p_{wb} - p_a) = k(T_a - T_{wb})(1 + 0.00115(T_a - T_{wb})),$$

k is a constant which depends on the atmospheric pressure, and the continuous line in Fig. 2.1 is appropriate for a normal sea level pressure of 101.3 kPa (Hodgman, 1965). The equation shows that the relation is not strictly linear, but that if $(T_a - T_{wb})$ is not more than 20 degC the slope will not change by more than about 2 per cent from one end to the other. The line in Fig. 2.1 therefore appears straight.

The sensible heat taken up by the wet bulb from the air is proportional to $(T_a - T_{wb})$, so the straight line relation implies that the evaporative heat loss must be proportional to $(p_{wb} - p_a)$. Since the rate of mass transfer (evaporation) is proportional to $(p_{wb} - p_a)$, the heat taken up per unit mass of water evaporated must be constant and equal to the latent heat of vaporization. (The total heat of evaporation approaches this value at high ambient humidities.) The dashed line shows the relation which would be expected at an air temperature of 35 °C if the heat taken up from the wet bulb were equal to the total heat of evaporation (Hardy, 1949). The total heat of evaporation is greater at low ambient humidities (large wet bulb depressions), so that the mass rate of evaporation, and therefore the vapour pressure difference, required to maintain the wet bulb in equilibrium on this assumption is less. The significant difference between the dashed line and the continuous line in Fig. 2.1 suggests that in the case of the wet bulb thermometer the heat removed from the bulb is equal to the latent heat of vaporization only, and that the remainder of the total heat of evaporation is derived from the environment. This conclusion relates to a fully wet surface from which evaporation occurs at a rate determined by the vapour pressure difference and the evaporation coefficient, i.e. at the maximum rate possible under the prevailing conditions.

The same may not be true if the rate of evaporation is less than the maximum, being limited by the sweat rate. Here Buettner's concept of relative humidity of the skin (Buettner, 1934) suggests that the sweat may be regarded as evaporating directly into a layer of air at skin temperature and at vapour pressure $\phi_s.p_s$, where ϕ_s is the relative humidity of the skin. If so, it would be proper to add to the latent heat of vaporization the heat taken up in expanding the vapour isothermally from p_s to $\phi_s.p_s$. This is equal to $RT_s\ln(1/\phi_s)$ (T_s is here in °K). Some examples are given in Table 2.3. Fig. 2.2 shows the relation between ϕ_s and λ_s. The effect of the heat of expansion on the total heat removed from the skin, λ_s, is considerable at low values of ϕ_s, and indeed approaches infinity at $\phi_s = 0$. Happily the rate of evaporation must then be zero, so this is not unrealistic. In the case of fully wetted skin, $\phi_s = 1.0$, there is no expansion of the saturated vapour at the skin surface, and the example of the wet bulb thermometer suggests

TABLE 2.3. *Total heat removed from the skin by evaporation, λ_s, (J/g) at various values of skin temperature, T_s, and skin relative humidity, ϕ_s. At $\phi_s = 1.0$ the total heat of evaporation equals the latent heat of vaporization. At other values of ϕ_s it is greater because of the heat taken up in isothermal expansion of the water vapour from saturation at skin temperature, p_s, to the vapour pressure at the skin surface, $\phi_s \cdot p_s$.*

T_s °C	ϕ_s					
	0.1	0.2	0.4	0.6	0.8	1.0
30	2751	2651	2554	2498	2457	2426
32	2748	2648	2550	2493	2452	2421
34	2745	2645	2546	2489	2448	2416
36	2743	2642	2543	2485	2444	2412
38	2740	2640	2540	2482	2440	2408

At $\lambda_s = 2400$, evaporation of 1.5 g/h gives a heat loss of 1.0 W.

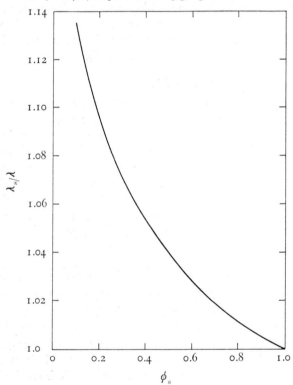

Fig. 2.2. The effect of skin relative humidity, ϕ_s, on the heat removed from the skin by evaporation, expressed here as the ratio of the total heat removed, λ_s, to the latent heat of vaporization, λ. When the skin is wet ($\phi_s = 1.0$) the total heat is equal to the heat of vaporization. At lower values of ϕ_s it is greater because of the heat taken up in expanding the water vapour to $\phi_s \cdot p_s$.

2. Convection and evaporation

that the subsequent expansion to ambient partial pressure takes up heat from the environment rather than from the skin.

The heat taken up in expanding the water vapour to ambient partial pressure (rather than only to $\phi_s p_s$) is infinite if the ambient air is completely dry, and the fact that evaporative cooling does not increase dramatically at very low ambient humidities provides further support for the view that the heat involved in expanding the vapour from $\phi_s . p_s$ to p_a is not taken up from the skin.

On the other hand one would expect a considerable increase in λ_s at very low atmospheric pressures. The rate of evaporation is $h_e(\phi_s . p_s - p_a)$ (see p. 38) and in dry air $\phi_s = E/h_e . p_s$. The evaporation coefficient, h_e, increases at low pressures, so for a given rate of evaporation ϕ_s diminishes, approaching zero in a (dry) vacuum. Water is lost through the skin even when the sweat glands are inactive, and might cause excessive cooling of astronauts wearing pressure suits of the leotard type (Webb, 1968).

Calculations involving $\dot{m}$ and λ_s

Equation (13) describes the mass rate of evaporation, $\dot{m}$, from a completely wet surface. For such a surface ϕ_s is 1.0, and the rate of heat loss can be obtained for this case from $E = \dot{m}.\lambda$. If the rate of sweat production is insufficient to maintain a film of water all over the skin surface, the vapour pressure at the skin surface, $\phi_s . p_s$, is the driving force for the diffusion of water vapour, which proceeds at a rate proportional to $(\phi_s . p_s - p_a)$ (see p. 38). This describes the mass rate of evaporation. The rate at which heat is taken up from the skin is $\dot{m}.\lambda_s$, or $\dot{m}(\lambda + RT\ln(1/\phi_s))$. Hence the general equation for evaporative heat loss is

$$E = h_e(\phi_s . p_s - p_a) (1 + (RT/\lambda)\ln(1/\phi_s)). \qquad (14)$$

Unless ϕ_s is known, solution of this equation is rather tedious, and the accuracy required in practice does not justify the labour. Thus, suppose we wish to calculate the mass sweat rate $(g/m^2.s)$ required to provide a certain rate of evaporative heat loss, E_{req}, under given conditions of skin temperature, ambient humidity and air movement. We first make an approximate estimate of ϕ_s from the incorrect equation, $E_{req} = h_e(\phi_s . p_s - p_a)$. Using this estimate, the appropriate value of λ_s is found from Table 2.3. The required mass sweat rate is approximately E_{req}/λ_s. The error involved is less than 0.5 per cent so long as ϕ_s is not less than 0.2.

Experimental determinations of λ_s on human subjects have given values somewhat greater than the latent heat of vaporization (Snellen, 1966; Belding, Hertig & Kraning, 1966; Mitchell *et al.* 1968). Snellen, Mitchell & Wyndham (1970) report a mean value of 2.60 kJ/g at ambient vapour pressures between 1.7 and 4.2 kPa. However, the regression equation for their pooled results was

$$E = 2469\dot{m} + 8.23,$$

where $\dot{m}$ is the rate of evaporation in g/s. The value 2.47 kJ/g is consistent with a skin relative humidity of about 0.7, close to the likely mean for the whole series. The constant term in the regression equation may represent a systematic error in calculated heat production or storage, but since it was not statistically significant the value of $\lambda_s = 2.60$ kJ/g, obtained by regression through the origin, was taken by the authors as the correct one.

The effect of salt on evaporation

Sweat contains salt, and when the water evaporates the salt is left behind on the skin. In hot dry climates with adequate air movement enough salt soon accumulates to saturate new sweat as it arrives on the surface. This has the advantage of causing the sweat to wet the skin more readily, forming a slimy film rather than tending to remain in droplets, but it has the disadvantage of reducing the vapour pressure. The ratio of the vapour pressure of a salt solution to that of water at the same temperature, the relative humidity of the solution, is almost independent of temperature, and it is convenient to express the depression of vapour pressure in this way. Values for sodium chloride are shown in Table 2.4. The depression of vapour pressure is by no means negligible, and in suitable conditions may greatly decrease the evaporative driving pressure. However, should the decrease in vapour pressure be such that evaporation is limited to less than the rate of sweat production, sweat will accumulate and drip off, washing the extra salt away with it, and the skin vapour pressure will increase (Macpherson & Newling, 1954).

While sodium chloride is the main constituent of sweat, many other substances are present in significant quantities (Robinson & Robinson, 1954). The depression of vapour pressure of mixtures of this sort cannot be calculated by merely summing the effects of the individual constituents. Measurements of the freezing point

2. Convection and evaporation

of sweat show that the depression is less than would be expected by summing the osmolarities of the constituents (Lichton, 1957; Adams, Johnson & Sargent, 1958), but close to that of mixtures of the most important constituents in appropriate concentrations (Foster, 1961). There is no biological magic about the effect of the constituents of sweat on its vapour pressure. Salt is the main constituent and the figures in Table 2.4 can be used as a rough guide to the behaviour of sweat.

TABLE 2.4. *Relative humidity of solutions of sodium chloride. The solubility of sodium chloride varies little with temperature, and a saturated solution contains about 36 per cent w/v salt.*

Salt concentration						
% saturated solution	0	20	40	60	80	100
Salt concentration % w/v	0	7	14	22	29	36
Relative humidity %	100	98	92	87	81	75

Under conditions of restricted evaporation the build-up of salt concentration on the skin surface is unlikely to reach levels of practical importance. Thus if the concentration of salt in secreted sweat is 0.5 per cent, and 10 per cent of the sweat is lost by dripping, the equilibrium concentration of salt on the skin surface will be 5 per cent, which has little effect on vapour pressure (Table 2.4). It is possible, however, that in long exposures sweat production may be reduced by hidromeiosis (p. 145) to such a degree that significant build-up of salt concentration on the skin surface may occur.

The effect of dissolved constituents on the latent heat of evaporation of sweat is probably negligible. The latent heat of vaporization and vapour pressure of a solution are related by the equation

$$\lambda = RT^2 \frac{d\ln p}{dT}.$$

Within the accuracy of the International Critical Tables, the relative humidity of strong sodium chloride solution is independent of temperature in the range 20–60 °C. Hence $d\ln p/dT$ for the solution is the same as for water, and the latent heats of vaporization must be equal.

Mean heat exchange coefficients for human subjects

For a human subject of given size, in air of fixed properties, the mean convection coefficient, $\bar{h}_c$, may be expected to vary approximately as a power of the air velocity, V.

$$\bar{h}_c = B \cdot V^n. \tag{15}$$

As n can have any fractional value, this equation is dimensional nonsense, but subject to the above limitations about the subject and the fluid it is an adequate practical representation of the relation more correctly described in equation (1). The velocity term does not specify the nature of the air movement (e.g. uniform wind, omni-directional room ventilation), the posture of the subject or his orientation to the air movement. Different values of B and n are to be expected for each type of situation so defined, and it may not be possible to read across from one to another. In the case of a uniform wind the velocity may be defined very simply as the wind speed, but the posture and orientation of the subject should also be specified. Where other types of air movement are present a single measure of air speed does not constitute an adequate description of the air velocity. It is usual in such cases to measure the air movement by a hot-wire or ion anemometer which is non-directional, and to calibrate this instrument in a wind tunnel so that its reading may be expressed as equivalent wind speed. The sensing element is very much smaller than a man, and it is a different shape, so it does not follow that if the anemometer reads the same in two environments having different types of air movement the convection coefficients for a human subject will be the same in these environments. Close agreement between different experimental determinations of the relation between $\bar{h}_c$ and V is not to be expected, both for this reason and because of the considerable practical difficulties involved in measuring the convective heat exchange and the skin temperature.

Carroll & Visser (1966) have shown that in a well-controlled wind tunnel it is possible to measure the convective heat exchange of a subject directly. Resistance thermometer grids extended across the wind tunnel section up and down stream of the subject sense the change in air temperature. If the velocity in the wake of the subject were uniform, as it is upstream, the heat exchange could be calculated directly, but unfortunately this is not so. It has been

2. Convection and evaporation

found, however, that heat emanating from a source within the zone immediately behind the body is distributed across the wake in the same way as heat originating in the body. Such a source can be used to produce known increments in the total heat taken up by the air, and the incremental changes in temperature difference between the grids can be used to provide an absolute calibration. An accuracy of about ± 5 per cent over the wind range 0.51 to 2.0 m/s has been obtained using a man-shaped model as the primary heat source. The method has been used with human subjects to determine the values of B and n in equation (15), which are shown in Table 2.5 (Mitchell *et al.* 1969).

Other determinations of $\bar{h}_c$ have depended on heat balance experiments in which the convective heat loss is calculated as the difference between the metabolic heat production and the heat loss by all channels except convection. The total sensible heat loss is found by subtracting from the metabolic heat production the evaporative heat loss (found by weighing) and the change in total body heat over the period of observation. This heat storage cannot be measured directly, and the estimate, based on changes in skin and deep temperatures, may be an important source of error (Gagge, 1936; Hardy & Dubois, 1938). Winslow, Gagge & Herrington (1940) found that the calculated value of $\bar{h}_c$ changed slowly over six hours towards what appeared to be an asymptotic value. Aikas & Piironen (1963) found that their estimates of $\bar{h}_c$ depended on whether the heat storage was calculated from the oesophageal or from the rectal temperature. Values based on the latter were some 40 per cent lower. Stolwijk & Hardy (1966a) were able to calculate storage satisfactorily for moderately warm conditions by weighting skin and tympanic temperatures, but were unable to obtain satisfactory results when the subject was cold.

If the sensible heat exchange is made large, so as to increase the accuracy of measurement, there are usually large differences in skin temperature from place to place, and the proper expression of mean skin temperature becomes difficult. Measurement of skin temperature at any one site becomes less accurate when the heat transfer is large (Molnar & Rosenbaum, 1963) and the tendency for the skin temperature to be lower in regions where the local value of h_c is large may affect the estimate of $\bar{h}_c$, since the total convective heat exchange is less than it would be if all the skin regions were at the mean skin temperature (Kerslake, 1963).

The direct determination of $\bar{h}_e$ is based on measurement of the rate of evaporation from a subject whose skin is completely wet with sweat. In order to maintain this state, the total sweat rate must considerably exceed the rate of evaporation, the subject is severely stressed and his skin temperature is usually rising. The heat storage does not matter, since the rate of evaporation is measured by weighing, but the changing skin temperature implies that the skin vapour pressure is changing, and the period of observation must be kept short. The ambient water vapour pressure has to be rather high in order to ensure that evaporative capacity does not exceed sweat production in the best ventilated parts of the body, and this means that the vapour pressure difference between skin and air is small. Sensible heat exchange can be kept small, increasing the accuracy of skin temperature measurement, but the other requirements considerably limit the accuracy with which $\bar{h}_e$ can be estimated.

TABLE 2.5. *Formulae for mean values of $\bar{h}_c$ and $\bar{h}_e$ for nude human subjects according to various investigators. The full references are given in the legend to Fig. 2.3, which shows the equations graphically.*

Source	Range of V	Formula	Posture
Buettner, 1934	0.15–0.50 m/s	$\bar{h}_c = 7.3\ V^{0.5}$ W/m². °C	Lying on bed
Gagge, 1937	0.03–0.51	$\bar{h}_c = 10.6\ V^{0.5}$	Reclining on net
Hardy & DuBois, 1938	'Still air'	$\bar{h}_c = 0.9$ to 1.6	Lying on net
Winslow *et al.* 1940	0.05–0.11	$\bar{h}_v = 11.6\ V^{0.5}$	Reclining on net
Nelson *et al.* 1947	0.15–3.00	$\bar{h}_c = 8.6\ V^{0.5}$ or $\bar{h}_c = 7.1\ V^{0.62}$	Standing Standing
Aikas & Piironen, 1963	0.15–0.22	$\bar{h}_c = 7.9\ V^{0.5}$	Lying on bed
Colin & Houdas, 1967	0.20–1.20	$\bar{h}_c = 2.7 + 8.7\ V^{0.67}$	Reclining on net
Mitchell *et al.* 1969	0.51–5.10	$\bar{h}_c = 7.2\ V^{0.60}$	Standing
Buettner, 1934	0.15–0.50	$\bar{h}_e = 120\ V^{0.5}$ W/m². kPa	Lying on bed
Nelson *et al.* 1947	0.15–3.00	$\bar{h}_e = 88\ V^{0.37}$	Standing
Clifford *et al.* 1959	0.58–4.00	$\bar{h}_e = 109\ V^{0.63}$	Standing

Formulae used by various investigators to describe their determinations of $\bar{h}_c$ and $\bar{h}_e$ are shown in Table 2.5, and represented

2. Convection and evaporation

graphically in Fig. 2.3. The scale for $\bar{h}_e$ in Fig. 2.3 has been adjusted to fifteen times that for $\bar{h}_c$. Theoretically one would expect the ratio $\bar{h}_e/\bar{h}_c$ to be 16.5 (Rapp, 1970). The lower value used here leads to a better fit between the lines for evaporation (9 and 11) and convection (5 a, 5 b and 8). It is also consistent with the observations of Brebner, Kerslake & Waddell (1958 a) on human subjects,

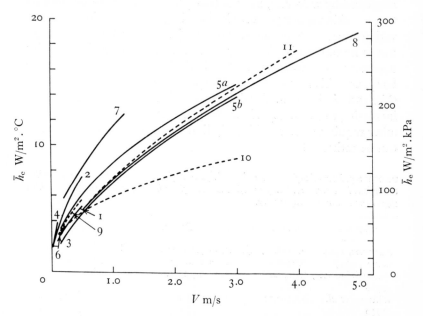

Fig. 2.3. Relations between wind speed and mean convection coefficient, $\bar{h}_c$, (continuous lines) and mean evaporation coefficient, $\bar{h}_e$, (dashed lines) for nude subjects according to various investigators (see Table 2.5). The scale for $\bar{h}_c$ is fifteen times that for $\bar{h}_e$ (see text, and cf. slope of straight line in Fig. 2.1). The sources are: 1, Buettner, 1934; 2, Gagge, Herrington & Winslow, 1937; 3, Hardy & DuBois, 1938; 4, Winslow, Gagge & Herrington, 1940; 5, Nelson et al. 1947; 6, Aikas & Piironen, 1963; 7, Colin & Houdas, 1967; 8, Mitchell et al. 1969; 9, Buettner, 1934; 10, Nelson et al. 1947; 11, Clifford, Kerslake & Waddell, 1959.

and with the behaviour of the wet bulb thermometer (Hodgman, 1965). Apart from lines 7 and 10 the formulae are in fair agreement with one another. In the case of line 7 (Colin & Houdas, 1967), the air velocity was measured near the skin, where it would be less than in the undisturbed stream. Values of $\bar{h}_c$ are therefore high. Line 10 represents the findings for $\bar{h}_e$ by Nelson et al. (1947).

Below about 0.5 m/s this line agrees with the others, but at higher wind speeds the values of $\bar{h}_e$ appear to be too low. The authors recognized the discrepancy between their findings for $\bar{h}_e$ and $\bar{h}_c$ and attributed it to failure to achieve full wetness at the higher wind speeds. In some cases as little as 10 per cent of the sweat was lost by dripping, and later work suggests that the better ventilated regions of the body would not have been fully wet under such conditions (Kerslake, 1963).

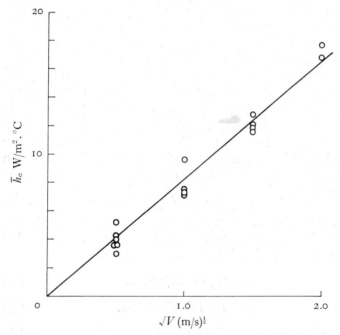

Fig. 2.4. Relation between convection coefficient, $\bar{h}_c$, and the square root of wind velocity. Points are taken from the lines in Fig. 2.3, omitting lines 7 and 10 (see text). Where $\bar{h}_e$ was measured, the calculated value of $\bar{h}_c$ (one fifteenth of $\bar{h}_e$) has been plotted here.

Some workers have determined the best fit values of both B and n in equation (15), while others have assumed a value of $n = 0.5$. There is no physical reason why the heat exchange coefficients should vary as the square root of the air velocity, but it is certainly easier to apply an equation containing the square root than one containing some other fractional power, and unless the accuracy

2. Convection and evaporation

of measurement precludes its use the square root formulation is to be preferred. In some cases the best fit value of n was significantly different from 0.5, but it is the spread of results between different investigators rather than the reproducibility within one experimental series which must be considered, since the application of formulae of this type is essentially to situations in which direct measurements cannot be or have not been made.

In Fig. 2.4 points from the curves in Fig. 2.3 are plotted against $\sqrt{V}$. Lines 7 and 10 in Fig. 2.3 have been omitted for the reasons given above, but otherwise a point from each line has been taken at each of the selected air speeds. Determinations at air speeds below 0.25 m/s are not included, but reference to Fig. 2.3 suggests that they would not be inconsistent with the others. The line, $\bar{h}_c = 8.28\sqrt{V}$ is a fairly good fit, although values at $V = 1.0$ are somewhat overestimated, and those at high velocities underestimated. There is no doubt that an equation based on $V^{0.6}$ would provide a better fit for all the observations, but in view of the scatter between the results of different workers and the greater convenience of the square root relation, the regression line shown in Fig. 2.4 may be regarded as an adequate compromise. Where greater accuracy is needed, direct determinations must be made in the environment in question, since possible differences in the nature of the air movement preclude the use of results obtained in another situation, however accurate they may be. For practical purposes the following formulae are recommended.

$$\bar{h}_c = 8.3\sqrt{V} \ \text{W/m}^2.°\text{C}, \tag{16}$$

$$\bar{h}_e = 124\sqrt{V} \ \text{W/m}^2.\text{kPa}. \tag{17}$$

Values of $\bar{h}_c$ and $\bar{h}_e$ at different values of V are tabulated in Appendix 4.

In using equations containing fractional powers of V one must be careful about approximation. For example, the expression for $\bar{h}_e$ given by Nelson et al. (1947) was $1.44V^{0.37}$, $\bar{h}_e$ being in kcal/m².h.mmHg, and V in ft/min. At $V = 100$ the value of $\bar{h}_e$ calculated from this equation is 7.9 kcal/m².h.mmHg. If the original equation is approximated by taking $V^{0.40}$, leaving the co-efficient 1.44 unchanged, the calculated value of $\bar{h}_e$ at 100 ft/min becomes 9.1 kcal/m².h.mmHg. According to the original formula this would correspond to a wind speed of 145 ft/min.

Fluctuating winds

Outside the laboratory, air velocity is rarely constant from moment to moment. In order to assign a mean effective wind speed to such environments (for the purpose of calculating the mean convection coefficient), it would be incorrect to take the mean wind velocity. Instead the mean square root velocity should be used, and this can be squared so as to return to the dimensions of velocity. The result is the square mean root velocity (not the root mean square, a function which would be used if the convection coefficient depended on the square of the wind velocity).

A similar consideration applies to the movement of limbs in walking. The velocity of air passing the legs varies during each step cycle, and the mean convection coefficient depends on the square mean root velocity, not on the mean velocity.

For clothed subjects the position is complicated by penetration of wind through clothing fabrics and ventilation through the clothing apertures (Chapter 5). These processes are not proportional to the square root of the wind speed, and require special treatment.

Submaximal evaporation

In many situations the rate of evaporation is not limited by the evaporative capacity but by the rate of sweat secretion. Clearly if the sweat rate is low and the air dry the sweat will evaporate as fast as it is produced, and it is the physiology rather than the physics which will determine the rate. It would nevertheless be useful to have an expression for evaporation in terms similar to those describing maximum evaporation; in particular to indicate the way in which ambient humidity impedes evaporation even if it does not limit it. Maximum evaporative capacity is given by

$$E_{\max} = \bar{h}_e(p_s - p_a). \tag{18}$$

Here p_s is the saturated water vapour pressure at skin temperature. If the actual rate of evaporation, E, is less than the maximum rate possible at the prevailing values of $\bar{h}_e$, p_s and p_a, some extra term must be introduced if E is to be expressed in this general form. In 1937, Gagge proposed the concept of wettedness. This is $E/E_{\max}$, the ratio of the actual rate of evaporation to the maximum

37

2. Convection and evaporation

under the prevailing conditions. If the whole of the skin were covered with a film of water the rate of evaporation would be E_{max} and the wettedness, W, would be 1.0. If the rate of evaporation were only half E_{max}, the value of W would be 0.5, and this rate of evaporation would be achieved if half the skin were covered with a film of water, the remainder being dry. One can think of the skin as covered by a mosaic of wet and dry patches of relative areas W and $(1 - W)$ respectively. Using this concept of wettedness, the equation for evaporation becomes

$$E = W.\bar{h}_e(p_s - p_a). \tag{19}$$

If E is constant while p_a varies, W will change. As ambient humidity increases, the skin surface becomes wetter, so increasing the area from which water is evaporating. It can be said that the wettedness is physiologically adjusted to provide the required rate of evaporation, but all that has happened is that the physiological factors which control sweat rate have been transferred to the new term W, where they are to some extent concealed.

An alternative approach proposed by Buettner in 1935 and later by Mole (1948) introduces the relative humidity of the skin ϕ_s. Retaining the symbols used above, the expression for evaporation is

$$E = \bar{h}_e(\phi_s.p_s - p_a). \tag{20}$$

Instead of regarding the skin as a patchwork of wet and dry areas, it is considered to have an effective vapour pressure, $\phi_s.p_s$, which is such that evaporation will proceed at the observed rate. This concept is subject to the same limitations as wettedness; both are merely ways of expressing the moisture concentration at the skin surface from which outward diffusion of water takes place (McLean, 1963).

The wettedness and the relative humidity of the skin can be related by equating the right-hand sides of equations (19) and (20).

$$W = (\phi_s.p_s - p_a)/(p_s - p_a), \tag{21}$$

$$\phi_s = W + (1 - W) p_a/p_s. \tag{22}$$

Equation (22) may be interpreted in terms of a mosaic of wet and dry areas, the former occupying a proportion, W, of the total area. For zero evaporation from the dry areas, these must be in vapour equilibrium with the air, i.e. the skin vapour pressure must be equal to p_a. Thus the contributions to the total skin vapour pressure $\phi_s.p_s$, will be $W.p_s$ from the wet areas and $(1 - W)p_a$ from the dry.

38

Wettedness and skin relative humidity are equal when $p_a = 0$ and when the skin is completely wet ($\phi_s = W = 1.0$). In other cases, since p_a cannot be negative and W is less than unity, ϕ_s is always greater than W (equation (22)). If there were no evaporation from the skin, W would be zero (equation (19)), and ϕ_s would be equal to p_a/p_s (equation (22)). This situation does not occur in practice, because even if there is no sweating, water is still lost from the skin by diffusion. However when the rate of evaporation is small it remains true that W is usually small, whereas ϕ_s may be quite large. The relation between ϕ_s and W at various values of p_a/p_s is shown in Fig. 2.5.

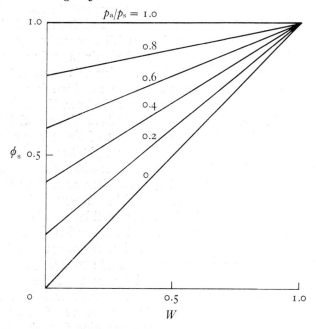

Fig. 2.5. Relation between the relative humidity of the skin, ϕ_s, and wettedness, W, at various values of p_a/p_s.

Regional evaporation

In the above account of the factors concerned in heat exchange between a subject and the surrounding air, the subject has been considered as a whole, and the heat exchange coefficients, skin temperatures and skin vapour pressures have been means for the

2. Convection and evaporation

whole man. Conditions of air movement vary from place to place over the skin surface, with consequent variations in the local values of h_c and h_e. Strictly, the local heat exchange coefficient should be applied to the local temperature or vapour pressure difference. The state of the air locally may be affected by heat exchanges elsewhere, but experimentally determined values of local heat exchange coefficients take this into account, and may be used with some confidence so long as there are not large differences in skin temperature or skin vapour pressure from place to place.

If all the sweat produced from all sites is evaporated, then either the local value of $h_e(p_s - p_a)$ must exceed the local sweat rate at all sites or the posture must be such that the unevaporated sweat runs down onto skin which is better ventilated and from which it evaporates. The reclining posture favours full evaporation by the latter method, since if the arms rest on the body the only places which drip are the buttocks and heels. Using this posture, the subject being supported on a net chair, the transition from full evaporation to fully wet skin can be abrupt (Brebner, Kerslake & Waddell, 1958 b). Fig. 2.6 shows results obtained with a reclining subject in an omnidirectional air movement of about 0.25 m/s. The air and wall temperatures were constant at 36 °C, and different humidities were used on different days. The results are means for the period 40–60 min after entering the environment. Over the humidity range 1.8 to 4.8 kPa the mean skin temperature is constant, as is the total evaporative loss. The small increase in sweat rate over this range balances decreased evaporation from the respiratory tract and decreased diffusion of water through the skin. At ambient humidities above 5.0 kPa both skin temperature and sweat rate increase steeply with humidity, some of the sweat dripping off. Measurements of $\bar{h}_e$ for the whole man in this situation showed that at a mean skin temperature of 36.1 °C, $E_{\max}$ would equal the required rate of evaporation from the skin if p_a were 4.9 kPa, indicated by the vertical line. At higher humidities, if evaporation were to proceed at the rate required for heat balance, the skin temperature would have to rise so as to maintain the driving pressure $(p_s - p_a)$. The calculated line is shown in Fig. 2.6, and the results suggest that under the conditions of these experiments the transition from evaporation limited by sweat production to evaporation limited by vapour pressure difference was abrupt.

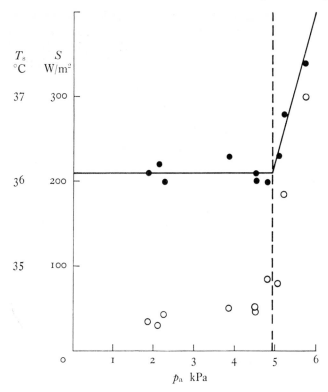

Fig. 2.6. Mean skin temperature, ●, and sweat rate, ○, of a resting subject at various ambient humidities. Air temperature 36 °C. The vertical dashed line at p_a = 4.9 kPa indicates the humidity at which the skin would become fully wetted. The continuous line indicates the mean skin temperature required for heat balance.

The case of a subject standing in a wind is quite different. With the subject facing into the wind the contrast between the dry chest and wet back, particularly between the shoulder blades, is obvious at quite moderate sweat rates in dry air. This is not due to a greater sweat rate on the back since if the subject faces down wind the back becomes dry. Measurement of local heat exchange coefficients is very difficult in Man, but presents no great problem in the case of models. The case of a vertical cylinder standing in a transverse wind had been examined in some detail, and conclusions about its behaviour can be applied with due caution to the human case (Kerslake, 1963).

2. Convection and evaporation

The ratio between the local evaporation coefficient, h_e, at various points around the circumference and the mean for the whole cylinder, $\bar{h}_e$, is shown in Fig. 2.7. The upwind position is at 0° and downwind at 180°. The diameter of the cylinder was 89 mm,

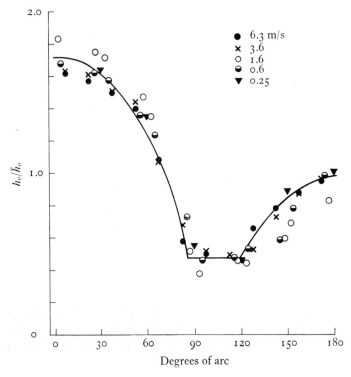

Fig. 2.7. Distribution of values of evaporative coefficient round the periphery of a cylinder in a transverse wind at various wind speeds. The ordinate is the ratio of the local value of h_e to the mean for the whole cylinder, $\bar{h}_e$, at the same wind speed. The abscissa shows the position on the circumference of the cylinder, 0° corresponding with the upstream position and 180° with the downstream position.

and the wind speed varied from 0.25 to 6.3 m/s ($N_{\text{Re}} = 1500–38000$). Although there were differences in the shapes of the curves at different wind speeds, the line drawn in Fig. 2.7 describes all wind speeds in this range fairly well and can be used to examine the likely behaviour of a uniformly sweating cylinder.

The maximum rate of evaporation from any point on the surface is $h_e(p_s - p_a)$, where h_e is the *local* coefficient for evaporation. If this is less than the sweat rate, S, (assumed to be the same at all points) that part of the surface will be fully wetted and the excess sweat will run off. The proportion of the area of the cylinder which

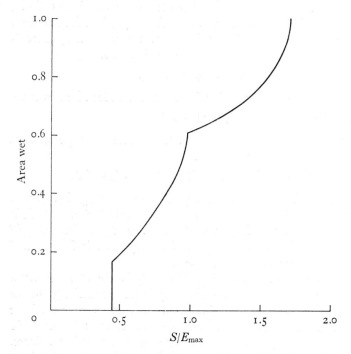

Fig. 2.8. The relation between the 'area wet' of a uniformly sweating cylinder in a transverse wind and the ratio between sweat rate and maximum evaporative capacity, S/E_{max}. The 'area wet' is the proportion of the total area over which the local sweat rate exceeds the local maximum evaporative capacity. The curve is derived from Fig. 2.7.

is fully wetted, which will be called the 'area wet', A_w, is that over which at all points $h_e < S/(p_s - p_a)$. For the whole cylinder, $E_{max} = \bar{h}_e(p_s - p_a)$, and therefore $(p_s - p_a) = E_{max}/\bar{h}_e$. Thus the condition for full wetting of the skin locally can be expressed

$$h_e/\bar{h}_e \leqq S/E_{max}.$$

The area wet, A_w, can be found by entering Fig. 2.7 at the level $h_e/\bar{h}_e = S/E_{max}$. Those parts of the curve below this level are those

43

2. Convection and evaporation

in which h_e is less than $S/(p_s - p_a)$ and where the surface is therefore completely wet. A_w is the number of degrees of arc thus delineated, divided by 180. Fig. 2.8 shows A_w as a function of S/E_{max}. The inflexions in this curve are due to the maxima and minima in the curve of $h_e/\bar{h}_e$ in Fig. 2.7.

The total rate of evaporation from the wet part of the cylinder may be found from the mean evaporation coefficient for that area. Calling the mean value of $h_e/\bar{h}_e$ for the wet part of the cylinder $\bar{n}_w$ and the total rate of evaporation from this part E_w,

$$E_w = A_w . \bar{n}_w . E_{max}.$$

The total rate of evaporation from the remainder of the cylinder E_d, is given by

$$E_d = (1 - A_w) . S$$

It is convenient to express both these rates of evaporation as proportions of the sweat rate, S.

$$E_w/S = A_w . \bar{n}_w . E_{max}/S, \quad E_d/S = (1 - A_w).$$

These and their total, E/S, can be plotted against E_{max}/S (Fig. 2.9). If less than 60 per cent of the sweat evaporates ($E/S < 0.6$), E_d is zero, the cylinder is fully wetted and $E = E_{max}$. When E_{max}/S exceeds 2.2, E_w is zero, no part of the cylinder is fully wetted and all the sweat evaporates, $E = S$.

The assumptions underlying this analysis are that the sweat rate and surface temperature of the cylinder are uniform. In the case of the human subject the important differences are that neither of these conditions is met and that the body is not cylindrical. To some extent it may be regarded as composed of cylindrical pieces, but these are of different diameters and would be expected to have different values of $\bar{h}_e$. The analogy must not be pressed, but the conclusion that the body does not become fully wetted until half the sweat is dripping off may be useful as a rough working guide. It is interesting that in the determinations of E_{max} by Clifford et al. (1959) the sweat rate was always at least twice the rate of evaporation, and the values obtained were in good agreement with expectation.

For the general case of a sweating object in which both sweat rate and heat exchange coefficients vary from place to place, a relation of the same general form is to be expected, i.e. that E/S will be some function of E_{max}/S. Any region of such an object will become fully wet if the local sweat rate per unit area exceeds the local maximum rate of evaporation. The local sweat rate and the local maximum rate of evaporation can be expressed as m and n times the respective means for the whole body. The condition for full local wetness is

$$m/n \geqq E_{\max}/S.$$

Each local region has its own value of m/n. As $E_{\max}/S$ decreases (e.g. as ambient humidity increases) the number of local regions

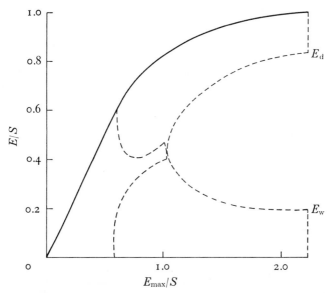

Fig. 2.9. The relation between E/S, the proportion of sweat which evaporates, and $E_{\max}/S$, for a uniformly sweating cylinder in a transverse wind. The dashed lines show the evaporation from fully wet parts of the cylinder (E_w) and incompletely wet parts (E_d). The continuous curve is the sum of these two. Up to $E/S = 0.6$ all parts of the surface are completely wet and $E/S = E_{\max}/S$.

in which m/n exceeds $E_{\max}/S$ will increase, so that the proportion of the total surface which is fully wet, A_w, will be a function of $E_{\max}/S$. The nature of this function need not concern us here; it is sufficient that A_w is determined by the fixed characteristics of the body and by $E_{\max}/S$. The mean values of m and n for the completely wet part of the body $\overline{m}_w$ and $\overline{n}_w$, will depend on A_w and so will also be functions of $E_{\max}/S$. As in the previous analysis, the rates of evaporation from the fully wet and incompletely wet regions of the body can be expressed in terms of the mean values of

2. Convection and evaporation

sweat rate, $\bar{m}_\mathrm{w}.S$ and evaporative capacity $\bar{n}_\mathrm{w}.E_\mathrm{max}$ for the wet region.

$$E_\mathrm{w}/S = A_\mathrm{w}.\bar{n}_\mathrm{w}.E_\mathrm{max}/S, \quad E_\mathrm{d}/S = (1 - \bar{m}_\mathrm{w}.A_\mathrm{w}).$$

The sum of these, E/S, is a function of A_w, $\bar{n}_\mathrm{w}$, $\bar{m}_\mathrm{w}$ and E_max/S. Since the first three of these are all functions of the fourth, it follows that
$$E/S = f(E_\mathrm{max}/S).$$

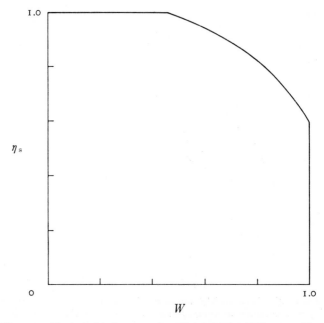

Fig. 2.10. The relation between the efficiency of sweating, η_s, (identical with E/S) and wetted area, W, for a uniformly sweating cylinder in a transverse wind.

The ratio E/S, the proportion of the sweat which evaporates from the skin, may be thought of as the efficiency of sweating, and assigned the symbol η_s. The wettedness, W, is E/E_max and η_s and W are related by the equation

$$\eta_\mathrm{s}/W = E_\mathrm{max}/S.$$

Since η_s is a function of E_max/S, it can also be expressed as a function of W, a form which has some practical convenience. This relation for the cylinder is shown in Fig. 2.10.

46

3 RADIATION

Heat radiation is a name for part of the electromagnetic radiation spectrum extending from the visible wavelengths (0.4–0.8 μm) to the much longer radio waves. There is no clear-cut distinction between heat radiation and radiation at other wavelengths – both visible light and radio waves liberate energy as heat when they are absorbed. The names merely relate to the most obvious manifestation of the radiation.

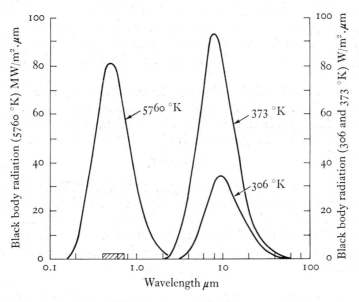

Fig. 3.1. Spectra of heat radiation from black bodies at 33 °C (306 °K), 100 °C (373 °K) and 5760 °K (solar temperature). The scale for the last is one million times that for the other two, and the tail of this spectrum is in fact above the other two curves. The wavelength of peak radiation at 100 °C is a little shorter (7.7 μm) than at 33 °C (9.4 μm) but the total radiation is more than twice as great. The scale for wavelength is logarithmic. The shaded box indicates the range of visible light.

3. *Radiation*

Although any part of the heat radiation spectrum may be present in industrial situations, natural environments usually involve two distinct bands, solar radiation, extending from the ultra-violet to the near infra-red, and radiation from surfaces below about 100 °C, in the far infra-red. The spectrum of heat radiation emitted from a surface depends on the temperature, the wavelength of the energy peak decreasing as the temperature increases. Representative spectra are shown in Fig. 3.1. The form of the spectrum is described by Wien's displacement law and the peak energy occurs at a wavelength λ^*, given by

$$\lambda^* = 2880/T \,\mu\text{m}, \tag{1}$$

where T is the absolute temperature (°K). The spectrum may thus be characterized by either λ^* or T. The appropriate value of T is known as the colour temperature. For the sun, λ^* is about 0.5 μm, so T is about 5760 °K. Absorption of some of the radiation by the atmosphere distorts the spectrum of sunlight at the earth's surface (Fig. 3.8), but the colour temperature generally provides an adequate working description of the whole spectrum. At 33 °C (306 °K), a representative temperature for human skin, λ^* is 9.4 μm.

The rate of emission of energy from a surface by radiation is proportional to the fourth power of the absolute temperature. The rate also depends on the nature of the surface, being greatest for a perfect black body, when it is given by the Stefan–Boltzmann equation

$$R = \sigma . T^4. \tag{2}$$

The constant, σ, has the value $5.67 . 10^{-8}$ W/m². °K⁴ (Hodgman, 1965). A surface which is not perfectly black (and this is the definition of blackness in this context) radiates less heat than this. The ratio of the actual emission of heat to that of a perfect black body at the same temperature is called the emittance, ϵ, so that for any surface

$$R = \epsilon . \sigma . T^4. \tag{3}$$

When radiant heat falls on a surface it may not all be absorbed. The proportion absorbed is called the absorptance of the surface, a black body absorbing all the radiation (absorptance 1.0) while other surfaces absorb some and reflect the remainder. The reflectance of a surface is the proportion of incident radiation which it reflects. Absorptance is equal to emittance, and reflectance to

48

$(1-\epsilon)$. These simple relationships are inevitable as the following proposition shows.

Consider the situation shown in Fig. 3.2 (a). Two surfaces are exchanging heat by radiation. One is assumed to have an emittance and absorptance of 1.0, the other having the mutually inconsistent values of 0.5 and 1.0 respectively. The temperatures of the two

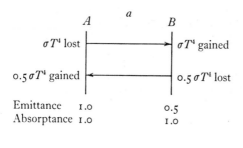

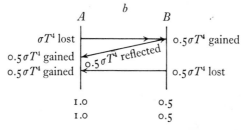

Fig. 3.2. Two surfaces exchanging heat by radiation. The surfaces are assumed to be at the same temperature, $T\,°$K. The emittance and absorptance of surface A are both 1.0. In diagram a the emittance of surface B is 0.5 and its absorptance 1.0. These inconsistent values lead to a net gain of heat by this surface. In diagram b both the emittance and absorptance of surface B are 0.5 and there is no net exchange of heat between the surfaces. Since there cannot be a net exchange of heat when the surface temperatures are the same, emittance must be equal to absorptance.

surfaces are equal. Surface A radiates at a rate $\sigma.T^4$, and all this energy is absorbed by surface B. Surface B radiates only $0.5\sigma.T^4$, all this being absorbed by surface A. Thus transfer of heat from A to B is greater than the transfer from B to A, a temperature difference will develop between the surfaces and this could be used to provide external work. Clearly this is absurd. In reality the rates of heat transfer in each direction must be equal. If surface B only radiates half as much energy as surface A it must absorb only

3. *Radiation*

half as much as surface A radiates, i.e. its absorptance must be 0.5, equal to its emittance (Fig. 3.2(b)). For a more precise account of absorption and reflection see Tregear (1966, p. 96).

Radiation exchange between surfaces

The equations for radiation exchange between surfaces can be derived from equation (3). The simplest case is that of two parallel surfaces of infinite size. Surface A radiates heat at a rate $\epsilon_A.\sigma.T_A^4$. When this radiation reaches surface B the proportion absorbed is ϵ_B, so the transfer from A to B is $\epsilon_B.\epsilon_A.\sigma.T_A^4$. In the same way the transfer from B to A is $\epsilon_A.\epsilon_B.\sigma.T_B^4$. Thus is might be supposed (wrongly) that the net radiation exchange would be the difference between these two.

$$R = \epsilon_A.\epsilon_B.\sigma(T_A^4 - T_B^4). \tag{4}$$

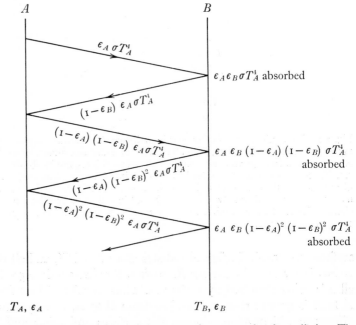

Fig. 3.3. The transfer of heat from one surface to another by radiation. The diagram shows the fate of radiation originating from surface A as it is reflected successively from one surface to the other. The total heat transfer is the sum of the quantities absorbed by surface B.

This equation is often quoted and applied, but it describes only the first passage of radiation between the two surfaces in each direction. It is correct if radiation is only reflected once (ϵ_A or $\epsilon_B = 1.0$). In other cases it is incorrect because the radiation which is not absorbed at the first passage is reflected back to the original surface where some is absorbed and the remainder again reflected. The quantities involved in successive reflections are indicated in Fig. 3.3 for radiation originating from surface A. At each collision with surface B the proportion absorbed is ϵ_B. The total transfer from A to B takes the form of a series describing the successive reflections.

$$R_{AB} = \epsilon_A . \epsilon_B . \sigma . T_A^4 (1 + (1-\epsilon_A)(1-\epsilon_B) + (1-\epsilon_A)^2 (1-\epsilon_B)^2 + \ldots).$$

Now the series, $1 + x + x^2 + x^3 \ldots$ equals $1/(1-x)$, so this equation can be expressed more simply

$$R_{AB} = \epsilon_A . \epsilon_B . \sigma . T_A^4 / (1 - (1-\epsilon_A)(1-\epsilon_B)).$$

Radiation originating from surface B behaves in the same way, and the net radiation transfer is the difference between the transfer from A to B and from B to A.

$$R = \epsilon_A . \epsilon_B . \sigma (T_A^4 - T_B^4) / (1 - (1-\epsilon_A)(1-\epsilon_B)).$$

With a little rearrangement this becomes

$$R = \sigma(T_A^4 - T_B^4) / (1/\epsilon_A + 1/\epsilon_B - 1). \tag{5}$$

When one of the emittances is unity this reduces to equation (4), and there is little difference between equations (4) and (5) if both emittances are high. Obviously the quantities of radiation involved in reflections will then be small and the fact that they are neglected by equation (4) will not be important.

Enclosures

In the case of an object of finite size in a relatively large enclosure the heat transfer by radiation does not follow equation (5). This is because radiation passing from the object to the enclosure and being reflected there does not all return to the object. Much of it will miss and strike another part of the enclosure. In consequence the emittance of the object is more important in determining the heat exchange than is that of the surrounding surfaces. The relations can be expressed exactly for certain simple shapes. For con-

3. Radiation

centric cylinders or spheres of surface areas A_1 (the smaller) and A_2, with emittances ϵ_1 and ϵ_2 respectively, the heat exchange per unit area of surface A_1 is given by

$$R = \epsilon_1.\sigma(T_1^4 - T_2^4)/(1 + \epsilon_1(1/\epsilon_2 - 1)\,A_1/A_2). \qquad (6)$$

For the case $A_1 = A_2$ (which implies that the enclosure is no larger than the object) this reduces to equation (5). The reflections would then occur in the same way as they do for infinite flat plates.

Equation (6), known as Christiansen's equation, can be used as an approximation for the radiant heat exchange between any body and any enclosure. Christiansen pointed out that its use should be restricted to cases in which the distance between the surfaces does not vary much from place to place and at least one of the surfaces reflects diffusely. It is noteworthy that in the case of a very large enclosure $(A_2 \gg A_1)$ the equation approaches

$$R = \epsilon_1.\sigma(T_1^4 - T_2^4). \qquad (7)$$

The emittance of a very large enclosure does not influence the radiant heat exchange at all. As an example of a moderate sized enclosure consider the case of a man of surface area 2 m² standing in a room 4 m × 4 m × 3m. Here A_1/A_2 is 1/40. If $\epsilon_1 = \epsilon_2 = 0.95$, equation (6) (the most correct one) leads to

$$R = \epsilon_1.\sigma(T_1^4 - T_2^4)/1.00125$$

which is very close to equation (7) but about 5 per cent different from equation (4). If ϵ_2, the emittance of the walls, is 0.20, equation (6) leads to

$$R = \epsilon_1.\sigma(T_1^4 - T_2^4)/1.095.$$

The effect of an 80 per cent reduction in emittance of the walls is to decrease the heat exchange by only 10 per cent. Silvering the walls of ones living room is not a very effective way of conserving body heat. Equation (4), which is sometimes erroneously applied to such a case, is in error by a factor of more than four, and its use could lead to some futile, though perhaps modish, styles of interior decoration. If the room is heated by infra-red radiation the economy may be greater since the subject now sees many images of the radiant source (Lorenzi, Herrington & Winslow, 1946; Bedford, 1964, p. 250.)

Radiation exchange between surfaces

All these equations for heat exchange by radiation contain the term $\sigma(T_1^4 - T_2^4)$, which is multiplied in each case by a quantity which depends on the emittances of the surfaces. It is convenient to refer to this quantity as an emittance factor, F_e. The general equation then becomes

$$R = F_e . \sigma(T_1^4 - T_2^4). \tag{8}$$

Computation of radiant heat exchange

The term $(T_1^4 - T_2^4)$ is rather tedious to evaluate, and the calculation can be eased by the use of Appendix 5. This shows values of $\sigma . T^4$ for various temperatures, here expressed in °C. Thus for surfaces at 50 °C and 30 °C, $\sigma(T_1^4 - T_2^4)$ is $618 - 478 = 140$ W/m². The radiant heat exchange in a particular case would be found by multiplying this by the appropriate emittance factor.

Configuration factors

In many environments the temperature of the surrounding surfaces is not uniform and it may be necessary to compute the radiation exchange between the skin and a surface such as that of a radiator panel. Reflections can be neglected unless the panel in question subtends a large solid angle at the skin surface. The radiant heat exchange is a function of the geometrical relation between the panel and the subject, which can be expressed by a configuration factor, F_c. This is applied in the same way as the emittance factor, F_e. Configuration factors for discs, spheres, cylinders and rectangles in various orientations are given by McGuire (1953). They have the useful property of being additive so that the factor for a rectangle with a circular hole in it is found by subtracting the factor for the hole from that for the rectangle. Factors for rectangles in relation to sitting and standing subjects are given by Fanger (1970, pp. 175–95).

First power relations

The dependence of radiant heat transfer on the difference between the fourth powers of the absolute temperatures of the surfaces raises some algebraic complications when heat exchange by radiation is combined with exchange by convection, since the latter depends on the difference between the first powers of the tempera-

53

3. Radiation

tures. If the temperatures are known there is no difficulty about computing the heat exchange, but if one requires to find the skin temperature consistent with a certain total heat exchange the algebra becomes cumbersome. It is often justifiable to use a first power approximation for the radiant heat exchange so that the general equation matches that for convection. The exact equation for radiant heat exchange may be summarized as

$$R = F_{ec}.\sigma(T_1^4 - T_2^4),$$

where F_{ec} is a factor including emittance and configuration terms. This can be re-expressed exactly as a first power relation

$$R = F_{ec}.h_r'(T_1 - T_2). \tag{9}$$

The first power radiation coefficient, h_r', analogous with h_C in the equation for convective heat exchange, itself depends on the temperatures of the two surfaces. (The symbol h_r' is used here because h_r conventionally indicates the first power coefficient including emittance and configuration factors).

$$h_r' = \sigma(T_1^3 + T_1^2.T_2 + T_1.T_2^2 + T_2^3).$$

The value of h_r' may be found from Appendix 5 if the temperatures are known. For example at 50 °C and 30 °C, h_r' is

$$(618 - 478)/(50 - 30) = 7.0 \text{ W/m}^2.°\text{C}.$$

If the two temperatures are equal the expression for h_r' reduces to $4\sigma.T^3$, which can be used for small temperature differences. The relation between h_r' and T_1 and T_2 is shown in Fig. 3.4.

It will be noticed that in order to find h_r' both temperatures must be known, and that in principle one cannot use the first power relation of equation (9) in the very case for which it is most desirable, when one of the temperatures is unknown. However the likely range of skin temperature is small, and when this is the unknown it may be possible to guess h_r' with an accuracy comparable with that with which other quantities, such as h_C, are known.

This first power relation may be somewhat misleading when the effect of changes in temperature of one of the surfaces is considered. Naïve differentiation of the equation $R = h_r'(T_s - \bar{T}_r)$, treating h_r' as a constant, gives $dR/dT_s = h_r'$. In the example cited earlier the mean radiant temperature, $\bar{T}_r$, was 50 °C, the skin temperature,

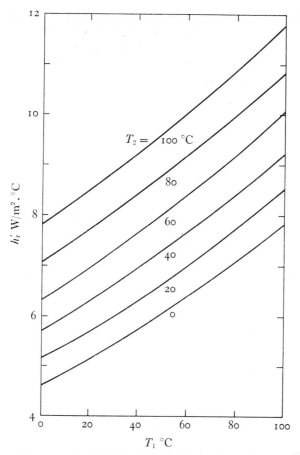

Fig. 3.4. The first power radiation coefficient, h_r' in equation (9), as a function of the temperatures, T_1 and T_2, of surfaces exchanging heat by radiation.

T_s, 30 °C and h_r' 7.0 W/m². °C. For this case, therefore, the inference would be that a change of 1 degC in skin temperature would change R by 7.0 W/m². This is wrong because the dependence of h_r' on T_s has been neglected. Differentiation of the algebraically identical equation $R = \sigma(T_s^4 - \overline{T}_r^4)$ gives $dR/dT_s = 4\sigma . T_s^3$, the value for h_r' when both temperatures are equal to T_s.

The reason why the change in R with T_s is independent of $\overline{T}_r$ can be appreciated by considering the basis of the radiant heat transfer process. The walls radiate at a rate depending on the wall

3. Radiation

temperature; the skin radiates at a rate depending on the skin temperature. Changing the skin temperature will affect the radiation leaving the skin but not that coming to it. We should therefore expect the effect of changing the skin temperature to be independent of wall temperature.

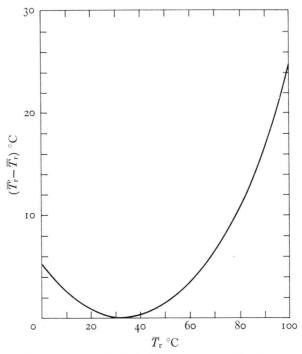

Fig. 3.5 The amount to be added to the true mean radiant temperature, $\bar{T}_r$, in order to obtain the effective mean radiant temperature, $\bar{T}'_r$, for use in equation (11).

A more satisfactory first power relation can be formulated for the case in which skin temperature is unknown. The relation between T_s and $\sigma \cdot T_s^4$ is nearly linear over the range 30–37 °C and can be represented within ± 1 W/m² by the equation

$$\sigma \cdot T_s^4 = 6.60(T_s - 30) + 478.$$

(The use of $(T_s - 30)$ allows adequate accuracy to be obtained with a slide rule.) The radiant heat exchange between the skin and the surrounding surfaces may be represented

$$R = F_{\epsilon c}(6.60(T_s - 30) + 478 - \sigma \cdot \bar{T}_r^4). \tag{10}$$

The last term can be found from Appendix 5. The constants in this equation are appropriate for the skin temperature range 30–37 °C. If the subject is clothed, the radiant heat exchange is between the environment and the clothing surface, the temperature of which may lie outside this range. For such cases the constants may require modification based on a guessed value of the clothing surface temperature.

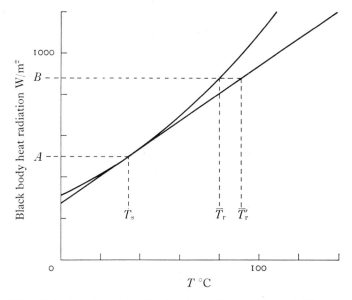

Fig. 3.6. Explanation of the effective mean radiant temperature. The curved line shows the relation between T and $\sigma . T^4$. The straight line is a tangent to it and the two lines coincide over the normal skin temperature range, 30–37 °C. At the indicated values of skin temperature, T_s, and true mean radiant temperature, $\bar{T}_r$, the radiant heat gain would be $F(B-A)$. If the heat exchange followed a first power law with $h'_r = 6.60$ (i.e. the straight line), the mean radiant temperature would have to be $\bar{T}'_r$ in order to obtain the same heat exchange. $\bar{T}'_r$ is equal to $\bar{T}_r$ for temperatures near 34 °C, but is higher in all other cases.

Although it provides a correct statement of the effect of changes in skin temperature, equation (10) does not have the simple form of the basic first power relation. In combining radiant and convective heat exchange it is desirable to retain the form of equation (9). In order to describe the effect of changes in skin temperature correctly, h'_r must be 6.60 W/m². °C (for the skin temperature

57

3. Radiation

range 30–37 °C). The required form of equation (9) can be retained by modifying the term for mean radiant temperature.

$$R = F_{ec} \cdot 6.60(T_s - \bar{T}'_r).$$ (11)

If $\bar{T}'_r$ were the true radiant temperature the equation would be incorrect for radiant temperatures outside the range 30–37 °C. $\bar{T}'_r$ must instead be an 'effective' radiant temperature, which can be found by equating the right-hand side of equation (11) with the true radiant heat exchange. For the skin temperature range 30–37 °C,

$$\bar{T}'_r = \sigma \cdot \bar{T}^4_r / 6.60 - 42.4.$$ (12)

Fig. 3.5 shows the amount to be added to $\bar{T}_r$ in order to obtain $\bar{T}'_r$. The rationale of this adjustment to $\bar{T}_r$ is demonstrated in Fig. 3.6.

Radiation area

In all the above equations heat exchange by radiation is expressed in watts per unit area. The area in question is that available for heat exchange by radiation and is less than the total surface area of the body because some parts 'see' others as part of the radiant environment. The radiant area factor behaves rather like a configuration factor for the whole man and may be conveniently expressed as the ratio of the effective area available for heat exchange by radiation (radiation area) to the total (DuBois) surface area. Use of this factor allows radiant heat exchange to be expressed per unit total surface area in accordance with the convention for other heat exchanges.

The radiation area depends on the posture, being greatest in the spread-eagle position and least when the subject is curled up. It can be determined by partitional calorimetry either by the rather unsatisfactory method of measuring total sensible heat loss at various air movements and extrapolating to zero convective heat loss, or better by varying the radiant temperature of the enclosure independently of air temperature and comparing the effect with the radiant heat exchange calculated for the whole DuBois area. Another approach is to photograph the subject from different directions, measure the projected areas and integrate over the whole sphere, i.e. over all possible directions (Guibert & Taylor, 1952). This is probably a more accurate method since it does not depend on measurement of body heat storage, and it has the further

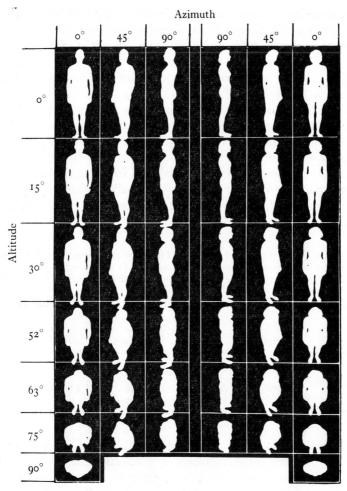

Fig. 3.7. Silhouettes of male and female subjects photographed from various angles of altitude and azimuth. (From Underwood & Ward, 1966.)

advantage that where the radiant temperature of the environment is not uniform (e.g. walking in the desert; hot ground, cool sky, sun) the effective areas may be computed for each component of the environment.

Silhouettes from various angles of altitude and azimuth obtained for subjects in the erect posture by Underwood & Ward (1966) using the photographic technique are shown in Fig. 3.7. The angle

3. Radiation

of altitude is zero when the subject is viewed from the horizon and 90° when he is viewed from directly overhead. The angle of azimuth is zero when he is viewed from directly in front and 90° when he is viewed from the side. Owing to the symmetry of the projected areas it is necessary to consider only angles between 0° and 90° in each case.

Projected areas were integrated by Guibert & Taylor (1952) for three postures and three male subjects of different build. The findings are summarized in Table 3.1 and are in general agreement with those of Underwood & Ward (1966) who found little difference between the area factors for male and female subjects. Fanger (1970, p. 170) obtained a somewhat smaller value (0.725) for standing subjects, but his figure for sitting subjects (0.696) is close to those in Table 3.1.

TABLE 3.1. *Ratios of radiation area to total (DuBois) area for subjects of different builds in different postures. The results were obtained by integrating projected areas obtained by photographing the subjects from many directions. (From Guibert & Taylor, 1952.)*

	Body build		
Posture	Heavy	Medium	Light
Erect	0.75	0.78	0.77
Semi-erect	—	0.72	—
Seated	0.72	0.70	0.69
Crouched	—	0.66	—

Halliday & Hugo (1962; 1963) have devised an ingenious method for measuring radiation area directly. An enclosure is uniformly illuminated with diffuse light and the luminous flux can be measured by observing the brightness of a small frosted window in one wall of the enclosure. When an object is introduced into the enclosure its surface absorbs some light and the brightness of the window diminishes. If the subject is coated with material of known absorptance for visible light his radiation area can be measured. Using models of men in the spread-eagle position, the radiation area was found to be about 95 per cent of the total area.

Full scale experiments with human subjects using this technique showed that the radiant area in the spread-eagle position was very nearly equal to the DuBois area (van Graan & Wyndham, 1964).

This suggested that the DuBois area underestimates the total surface area, as Banjeree & Sen (1955) had concluded from direct measurements of the surface area.

Solar radiation

Emittance depends on wavelength, and while in the long infra-red most non-metallic surfaces have emittances close to unity, in the visible range the values may be much lower. Black and white paints or skins behave roughly as black bodies in the long infra-red (Mitchell, Wyndham & Hodgson, 1967), but obviously reflect very differently in the visible range. Fig. 3.8 shows the spectrum of sunlight received at the earth's surface and the absorptances of white and Negro skin. The absorptance curves roughly mirror the solar spectrum; at wavelengths where the solar energy is high the absorptance is low. Over the whole spectrum the white skin absorbs about 60 per cent of the incident radiation, the Negro skin about 80 per cent. Pigmentation increases the heat load, but protects the deeper layers of the epidermis from damage by ultra-violet (0.3–0.4 μm).

Apart from certain industrial situations, human subjects are usually surrounded by surfaces radiating in the long infra-red, with the single exception of the sun. Solar radiation constitutes an important heat load which can be treated somewhat differently from the infra-red heat exchanges.

The sun appears as a rather small object occupying only about five millionths of the radiant environment. Its size can therefore be neglected in computing radiant heat exchange with the sky, and its contribution to the total heat load can be expressed simply as the additional heat absorbed by the skin or clothing. The direct rays of the sun fall on the projected area of the subject, which varies with the solar altitude and azimuth (Fig. 3.7). Projected areas for various angles are shown in Fig. 3.9. The solar energy falling directly on the subject is the product of his projected area and the intensity of the solar radiation flux. The skin or clothing reflects some of this radiation, absorbing the remainder. The total heat gain from direct sunlight is given by

$$H_{\mathrm{sd}} = A_{\mathrm{p}}.\alpha_{\mathrm{s}}.I_{\mathrm{s}}, \tag{13}$$

where A_{p} is the projected area for the prevailing solar angles, α_{s} the absorptance of the skin or clothing for sunlight (proportion of

3. *Radiation*

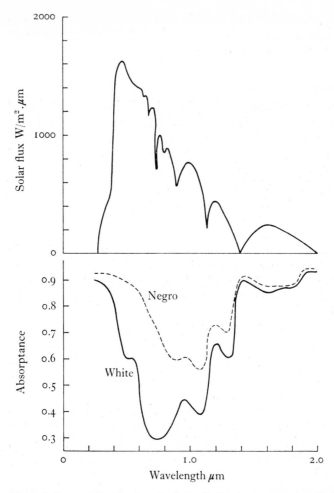

Fig. 3.8. Spectral distribution of solar radiation (upper curve) and of absorptances of white and Negro skin (lower curves). Redrawn from Gates, 1962 and Jacquez *et al.* 1955 *a, b*.

incident solar radiation absorbed) and I_s the solar radiation flux (W/m²) measured normal to the radiation. The fact that some parts of the subject face the sun squarely while the light only grazes others is unimportant. The projected area intercepts a certain quantity of sunlight, all of which strikes the subject and is either reflected or absorbed. Pugh & Chrenko (1966) have shown that for

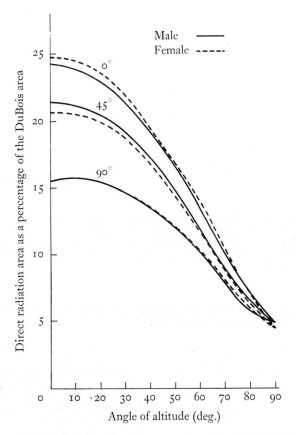

Fig. 3.9. Projected area, expressed as proportion of total (DuBois) area, of male and female subjects as a function of solar altitude. Curves are shown for three angles of azimuth (0° corresponds to facing the sun). (From Underwood & Ward, 1966.)

thickly clothed subjects the ratio of projected area to total surface area of clothing is substantially the same as the equivalent ratio for the nude subject (Fig. 3.10). Formulae for estimating the projected area of a standing subject are presented in Table 3.2.

Estimation of the solar heat load by this approach requires knowledge of the projected area and of the intensity of the solar radiation flux measured normal to the direction of the sun's rays. In field situations it is much easier to trace the shadow of the subject on a horizontal surface than to trace the projected area by

3. Radiation

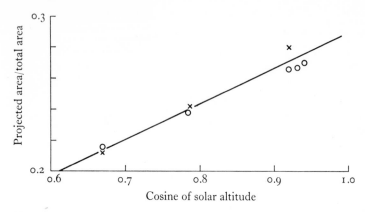

Fig. 3.10. Ratios of projected area to total surface area as a function of the cosine of the solar altitude for a nude subject (×) and the same subject dressed in thick clothing (○). In the former case the total surface area is taken as the DuBois area, in the latter as the total surface area of the clothing. (Observations by Chrenko & Pugh, 1961.)

TABLE 3.2. *Formulae for estimating the ratio of the projected area to the DuBois area for standing subjects. Solar altitude, α.*

Source	Basis	Formula	Orientation
Pugh & Chrenko, 1966	Stylised model	$0.33\cos\alpha + 0.038\sin\alpha$	Facing sun
		$0.17\cos\alpha + 0.039\sin\alpha$	Sideways
Underwood & Ward, 1966	Elliptical cylinder	$0.42\cos\alpha + 0.043\sin\alpha$	Facing sun
		$0.28\cos\alpha + 0.043\sin\alpha$	Sideways
Taylor, 1956	Cylinder	$0.29\cos\alpha + 0.039\sin\alpha$	
Lee, 1964		$0.30\cos\alpha + 0.039\sin\alpha$	
Pugh & Chrenko, 1966	Field observations $\alpha = 20\text{--}40°$	$0.23\cos\alpha + 0.06$	Facing sun

holding the surface so that it is normal to the sun's rays. The horizontal shadow is the horizontal area which is not reached by the solar radiation, so the total quantity of radiation intercepted by the subject must be the product of the shadow area and the solar intensity measured by means of a flat surfaced radiometer held horizontally (horizontal pyrheliometer) (Chrenko & Pugh, 1961; Lee, 1964; Lee & Vaughan, 1964). Not only is the technique of area measurement easier but the radiometer can be set up permanently to provide appropriate values throughout the day.

Indirect solar radiation

Sunlight may also reach the subject after reflection from the terrain or clouds, and scattered sunlight comes from all parts of a clear sky. Because of the different absorptance of the skin or clothing at different wavelengths it is necessary to distinguish this reflected sunlight from the infra-red radiation emitted by the surrounding surfaces. The intensity in any direction can be measured by comparing radiometer readings obtained with different filters. Representative values for direct and indirect solar radiation in various type of terrain are shown in Table 3.3.

TABLE 3.3. *Solar radiation flux in various types of terrain. The solar angle is taken as 60° in the first three cases and 45° for the subarctic region.* (From Roller & Goldman, 1968.)

| Terrain | Solar radiation flux W/m^2 | | |
	Direct	Reflected from sky	Reflected from terrain
Desert	1000	188	133
Rain Forest	840	180	46
Tropical steppe	1020	114	105
Subarctic	950	143	64

The partition of radiant heat exchange for clothed subjects marching in the Arizona desert is shown in Table 3.4. By far the largest component is direct solar radiation.

TABLE 3.4. *Partition of radiant heat exchange for clothed subjects marching in the desert. Radiant heat gain is expressed per unit DuBois area.* (From Lee, 1964.)

	Radiant heat gain W/m^2
Direct solar radiation	170
Reflected solar radiation from sky	10
Reflected solar radiation from terrain	28
Infra-red exchange with terrain	49
Intra-red exchange with sky	−34
Net gain	223

3. Radiation

Operative temperature

When a first power relation is used to describe radiant heat exchange the processes of convection and radiation have much in common. Each depends on the difference between skin temperature and an environmental temperature, T_a or $\bar{T}_r$, and on a heat exchange coefficient, h_c or h_r. Conventionally h_r includes emissivity and radiant area factors so that the total sensible heat exchange may be represented

$$(C+R) = h_c(T_s - T_a) + h_r(T_s - \bar{T}_r). \tag{14}$$

As a first step in the construction of systems of equivalent environments it would be useful to be able to reduce the four factors, T_a, $\bar{T}_r$, h_c and h_r to two by substituting an appropriately derived temperature and heat exchange coefficient which would provide the same rate of sensible heat transfer. This approach was developed at the Pierce Laboratory (Herrington, Winslow & Gagge, 1937), and the derived quantities are known as the Operative temperature, T_o, and the Operative heat exchange coefficient, h_o. We wish to represent the sensible heat transfer in an equation of the form

$$(C+R) = h_o(T_s - T_o). \tag{15}$$

By equating the right-hand sides of equations (14) and (15), expressions for T_o and h_o can be found.

$$T_o = (h_c \cdot T_a + h_r \cdot \bar{T}_r)/(h_c + h_r), \tag{16}$$

$$h_o = h_c + h_r. \tag{17}$$

It will be seen that T_o and h_o are independent of skin temperature, and that Operative temperature (a weighted mean of T_a and $\bar{T}_r$) appears to be a function of environmental factors only. This is not quite true because the values of h_c and h_r in a given environment depend on the shape and size of the subject as well as on the physical properties of the environment. (Emittances and geometrical relations are also involved in h_r.) However, the physiological responses of the subject are not involved in the calculation of T_o and h_o, and where size and shape can be assumed, the approach provides a concise description of the sensible heat stress of the environment. If the radiant temperature is not very different from skin temperature the simple first power relation of equation (9) may be used, but at high radiant temperatures it is sometimes better to use the effective radiant temperature (equation (12)). The Operative temperature is discussed further in Chapter 8.

The globe thermometer

Vernon (1932) showed that the mean radiant temperature of the environment could be measured by means of a globe thermometer consisting of a hollow sphere, blackened on the outside, with a thermometer bulb at the centre. He used a sphere 6 inches (150 mm) in diameter, because thin copper spheres of this size were readily obtainable, being used as floats in domestic water systems. The globe reaches thermal equilibrium when the heat gain by radiation equals the heat loss by convection. Using the first power approximation

$$h_{rg}(\bar{T}_r - T_g) = h_{cg}(T_g - T_a). \tag{18}$$

Here T_g is the equilibrium temperature of the globe and h_{rg} and h_{cg} the radiation and convection coefficients. The relation between h_{cg} and air movement was established experimentally, so that $\bar{T}_r$ could be found from measurements of globe temperature and air movement. If the globe temperature is in the range 30–37 °C, and h_{rg} is taken as 6.60, equation (18) leads directly to the effective mean radiant temperature, $\bar{T}_r'$.

Using the fourth power relation for radiation, the equation for the globe thermometer is

$$\sigma(\bar{T}_r^4 - T_g^4) = h_{cg}(T_g - T_a) \tag{19}$$

which may be solved for $\bar{T}_r$ by using Appendix 5. Thus, suppose h_{cg} is 10.0 W/m². °C, T_g is 35.0 °C and T_a 30.0 °C. The right-hand side of equation (19) becomes 10.0(35.0 − 30.0) = 50.0 W/m². °C. From Appendix 5, $\sigma . T_g^4$ is 510 W/m², so $\sigma . \bar{T}_r^4$ must be 560 W/m². The same table gives the approximate value of $\bar{T}_r$, 42.2 °C.

Bedford & Warner (1934) found that the globe temperature was well correlated with sensations of warmth, and appeared to provide a direct indication of sensible heat stress, at least for human subjects. Vernon's choice of a 6 inch globe was a fortunate one, since it happens that a sphere of this size weights air and radiant temperatures nearly correctly for a human subject. Equation (18) can be re-arranged to provide an expression for globe temperature.

$$T_g = (h_{cg}.T_a + h_{rg}.\bar{T}_r)/(h_{cg} + h_{rg}). \tag{20}$$

3-2

3. *Radiation*

This is strictly analogous with equation (16) for Operative temperature, and the globe would clearly provide a direct reading of Operative temperature if its convection and radiation coefficients were equal to those for the human subject. This would be hard to arrange, and is fortunately not the only condition which will provide the desired result. The weighting factors for air temperature in the Operative temperature and globe temperature equations are

Operative $h_c/(h_c + h_g)$,
Globe $h_{cg}/(h_{cg} + h_{rg})$.

If these weightings are the same the Operative and globe temperatures in any environment will be equal. The requirement for equal weighting is

$$h_c/h_{cg} = h_r/h_{rg}.$$

The factor for radiant area is included in the radiation coefficient, h_r, for the human subject. In the case of the globe the radiant area is equal to the total area, and if the emittances of the subject and the globe are equal the requirement can be expressed

$$h_c/h_{cg} = A_r/A_D,$$

where A_D is the DuBois area of the subject and A_r his radiation area. The ratio A_r/A_D is less than unity, so the convection coefficient for the globe must be larger than that for the man. This allows the globe to be smaller than it would need to be if h_{rg} were equal to h_r.

According to Williams (McAdams, 1942, p. 237), the convective heat exchange of a sphere in a wind is described by

$$N_{Nu} = 0.33 N_{Re}^{0.6}. \tag{21}$$

In air at 20 °C the expression for the convection coefficient of a 150 mm globe is

$$h_{cg} = 14.1 V^{0.6}.$$

At $V = 0.25$ m/s, h_{cg} is 6.1 W/m^2. °C. The convection coefficient for a human subject at this wind speed is about 4.2 W/m^2. °C (Fig. 2.3). The ratio h_c/h_{cg} is thus about 0.69, close to the value of A_r/A_D for a seated human subject (Table 2.1). The globe therefore provides a direct reading of Operative temperature appropriate for office and domestic situations.

For standing subjects the globe underestimates the effect of radiant temperature. Mitchell *et al.* (1969) found that the convection coefficient for standing nude subjects was $7.2V^{0.6}$ W/m². °C, slightly more than half that of the globe, while the ratio A_r/A_D for the standing subject is about 0.78 (Table 3.1). A larger globe would be more appropriate for this case (American Society of Heating and Ventilating Engineers, 1942).

Sunlight

Woodcock, Pratt & Breckenridge (1960) have examined the behaviour of the globe thermometer in sunshine. The heat gained directly from the sunlight per unit total area by a human subject or a globe is given by

$$H_{sd} = \alpha_s . I_s . A_p/A_D,$$

where A_p is the projected area, and A_D the total surface area. For a man the ratio A_p/A_D depends on posture and orientation, but for a sphere the ratio is constant at 0.25 (area of cross section πr^2, total area $4\pi r^2$). A roughly man-shaped globe thermometer would avoid difficulties about posture and time of day (Winslow & Greenberg, 1935) but Woodcock *et al.* (1960) have pointed out that for many practical cases the ratio A_p/A_D for a man can be taken as about 0.20 (cf. Fig. 3.9). If the globe were given the same value of α_s as the man by a suitable coating (Ambler, undated) its heat gain per unit area from the sunlight would be about 1.25 times that for the man. Its convection coefficient is about 1.4 times that for the man so that the temperature of such a globe should give an approximate indication of the Operative temperature. The scattered and reflected sunlight, which contribute to the heat load on the man, affect the globe to about the right extent. A globe covered with a sample of the outer clothing has been shown to give an appropriate indication of sensible heat stress on troops in the field (Yaglou & Minard, 1956).

Small globe thermometers

The 150 mm diameter globe, which gives an approximately correct direct reading of Operative temperature for the human subject, is sometimes inconveniently large for practical use. If a smaller globe is used, the reading of the standard instrument can be calculated if the air movement is known (Hey, 1968). The need for this measurement is not a serious disadvantage, since Operative temperature alone does not define the sensible heat stress of the en-

3. Radiation

vironment. The Operative heat transfer coefficient must also be known, and this depends on air movement.

The relation between wind speed and convection coefficient for a standard 150 mm diameter globe at normal atmospheric pressure is shown in Fig. 3.11. The effect of the size of a sphere on its convection coefficient can be found by expanding equation (21)

$$h_{\mathrm{cg}} = 0.33K.\nu^{-0.6}V^{0.6}.L^{-0.4}.$$

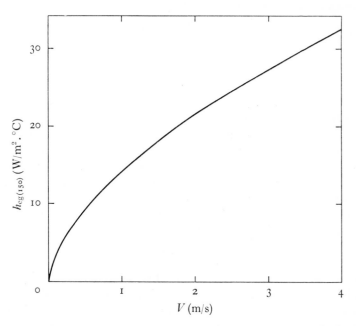

Fig. 3.11. Relation between convection coefficient, $h_{\mathrm{cg}(150)}$, of a 150 mm diameter globe and wind speed, V, at 30 °C and normal atmospheric pressure.

The convection coefficient varies inversely as the 0.4 power of the characteristic dimension, e.g. diameter. The effect of size is shown in Fig. 3.12, where a size factor is plotted against the diameter. The factor is arranged so that it has the value 1.0 for a 150 mm diameter globe. The convection coefficient for a globe of some other size can be found by multiplying the convection coefficient for the standard globe (Fig. 3.11) by the appropriate size factor.

Because a small globe has a larger convection coefficient than a large globe at the same air speed, its temperature will be closer

70

to air temperature and the accuracy with which mean radiant temperature can be measured will be less. Thus, taking the example cited earlier;

$$h_{cg(150)} = 10.0 \text{ W/m}^2.°\text{C}, \quad T_g = 35.0 °\text{C}, \quad T_a = 30.0 °\text{C},$$

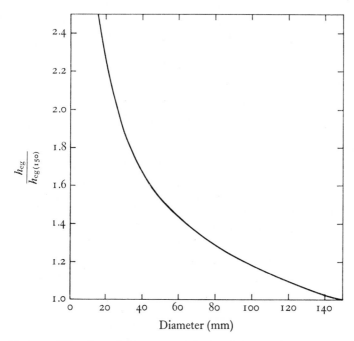

Fig. 3.12. The effect of size on convection coefficient. The abscissa shows the diameter of the globe, the ordinate the ratio of its convection coefficient, h_{cg}, to that of a standard 150 mm diameter globe, $h_{cg(150)}$.

the temperature of a 50 mm diameter globe can be calculated. Since T_g is in the range 30–37 °C, the effective radiant temperature can be found from

$$6.60(\bar{T}_r' - T_g) = h_{cg}(T_g - T_a)$$

$\bar{T}_r'$ comes to 42.6 °C (rather higher than the true radiant temperature). This value of $\bar{T}_r'$ can be substituted in the corresponding equation for the smaller globe ($h_{cg} = 15.3$ W/m^2.°C), the temperature of which comes to 33.8 °C. The temperature of a standard globe can be found from that of a small globe by a similar process.

3. Radiation

Indoors, when air movement is low and radiant temperature not very different from air temperature, the Pierce Laboratory workers have suggested the use of a two inch (50 mm) diameter globe as a direct indicator of Operative temperature. The amount by which the small globe temperature exceeds air temperature multiplied by 1.4 gives an approximate indication of the amount by which Operative or standard globe temperature exceeds air temperature. An instrument has been designed which will give a direct reading (Gagge *et al.* 1968).

Response time

The standard Vernon globe with a mercury thermometer at the centre takes 20 to 30 min to reach equilibrium. The process is a complicated one since three components must come into equilibrium, the metal of the globe itself, the air contained in it, and the thermometer. The heat capacities of these components are respectively about 150 J/°C, 2 J/°C, and 7 J/°C. The surface area is about 720 cm² and the total heat capacity about 2200 J/m², of which 94 per cent is due to the metal of the globe itself. If we neglect the heat taken up by the air and the thermometer, we should expect the globe to approach equilibrium with a time constant of $2200/h_0$ seconds. (The time constant is the factor, k, in the expression which describes the approach to equilibrium temperature in a simple system,

$$(T_\infty - T_t) = (T_\infty - T_{t0})\, e^{-t/k},$$

where t is the time, T_{t0} the temperature at time o, T_t the temperature at time t, and T_∞ the final equilibrium temperature.) At an air movement of 0.1 m/s, h_0 for the globe is about 10 W/m².°C so the time constant would be about 220 s (3.7 min). Equilibrium is closely approached (98 per cent complete) after four time constants, about 15 min. The time would be shorter at higher air movements. In practice the thermometer approaches equilibrium more slowly than this because of the low conductance between the globe and the thermometer.

This suggests that reducing the thickness of the metal would not have much effect on the overall speed of response, as Hellon & Crockford (1959) have demonstrated experimentally. They further showed that the response time could be reduced considerably by replacing the thermometer by a thermocouple. Reducing the mass of the sensing element has little effect on the total heat capacity but reduces the lag between the temperature of the globe and that of the sensor. A further small improvement was produced by stirring the air inside the globe, but its extent does not justify the use of a stirrer in routine measurements.

Effective Radiant Field[1]

Workers at the Pierce Laboratory, where the concept of Operative temperature originated, have proposed an alternative approach to the integration of radiant and convective heat exchange (Gagge, Stolwijk & Hardy, 1965; Gagge, Rapp & Hardy, 1967). As in the case of Operative temperature the actual situation is compared with an imaginary one in which mean radiant temperature and air temperature are equal, but in this case both are imagined to be equal to the actual air temperature. In the actual situation the sensible heat exchange is given by

$$(C+R) = h_c(T_s - T_a) + F_{\epsilon c} . \sigma(T_s^4 - \bar{T}_r^4).$$

In the imaginary situation, $\bar{T}_r = T_a$, an additional term is introduced, known as the Effective Radiant Field, H_r.

$$(C+R) = h_c(T_s - T_a) + F_{\epsilon c} . \sigma(T_s^4 - T_a^4) - H_r.$$

Equating the sensible heat transfer in the two cases,

$$H_r = F_{\epsilon c} . \sigma(\bar{T}_r^4 - T_a^4).$$

The Effective Radiant Field is the radiant heat gain that the subject would have if his skin were at air temperature. One does not have to know the skin temperature in order to find H_r, which can readily be evaluated from Appendix 5. The total Effective Radiant Field is the sum of the contributions of different surfaces in the environment. Surfaces at air temperature make zero contribution to H_r, and the contributions of others such as radiator panels, windows, etc. can be calculated in each case from the surface temperature, air temperature and configuration factor.

The effective Radiant Field is related to Operative temperature by the equation

$$T_0 = T_a + H_r/h_0. \tag{22}$$

[1] Also known as Effective Radiant Flux (Gagge, 1970).

3. Radiation

The coefficient, h_0, is defined by equation (17) as the sum of h_c and h_r, h_r being the first power radiation coefficient appropriate for exchange between the skin or clothing surface and the surroundings, not for a surface at air temperature. The distinction is unimportant unless the skin and air temperatures are very different.

The Operative temperature required for comfort can be specified for given conditions of clothing and activity. Air movement must be within a rather restricted range if stuffiness and draughts are to be avoided, and this in effect defines h_0. It is desirable that the mean radiant temperature should not be lower than air temperature (Bedford, 1964, p. 144), and with this proviso the engineer is free to manipulate H_r and T_a according to the means at his disposal. The method considerably simplifies the problem of optimizing heating and air-conditioning systems.

Radiation as a temperature increment

It is sometimes useful to regard environmental heat radiation as an equivalent increment in air temperature (thermal radiation increment). The situation of a man exposed to sunlight may be expressed as an equivalent shade temperature which, for the nude subject, is identical with the Operative temperature. (It is assumed that in the shade condition the surrounding surfaces will be at air temperature.) The thermal radiation increment is thus $(T_0 - T_a)$, which, from equation (22), is equal to H_r/h_0. Burton & Edholm (1955) have shown that experimental observations made by Robinson & Turrell (1944) on subjects wearing shorts are in fair agreement with the calculated expectation. Lee & Vaughan (1964) obtained somewhat lower increments, comparable with the results of Adolph & Molnar (1946).

The thermal radiation increment H_r/h_0 applies strictly to the case of the nude subject, or to sensible heat exchange at the surface of clothing which is opaque to radiation. The increment for clothed subjects, judged by their physiological state, is likely to be less if there is significant ventilation within the clothing or if sweat is able to evaporate directly from the skin through the clothing. It may be greater if the radiation penetrates the clothing. The effects of clothing are considered in more detail in Chapter 5.

4 THE HEAT BALANCE OF SWEATING SKIN

Wet bulb temperature

The relations between air temperature, water vapour pressure and wet bulb temperature can be represented on the psychrometric chart (Fig. 4.1) using the first two quantities as axes.

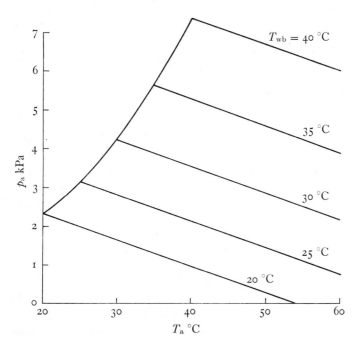

Fig. 4.1. The psychrometric chart. The axes are air temperature, T_a, and water vapour pressure, p_a. The curved line indicates saturated air. The wet bulb temperature is shown by the sloping straight lines, which are parallel but are not evenly spaced. At saturation wet bulb temperature equals air temperature. The chart may be extended by adding lines of equal relative humidity. (Humidity ratio is sometimes used as ordinate.)

4. *The heat balance of sweating skin*

The wet bulb temperature is the equilibrium temperature of an unheated wet surface exchanging heat by evaporation and convection only.

$$h_e(p_{wb} - p_a) = h_c(T_a - T_{wb}). \tag{1}$$

Here h_e is the evaporation coefficient, p_{wb} the saturated water vapour pressure at the wet bulb temperature (i.e. the water vapour pressure at the surface of the wet bulb), p_a the ambient water vapour pressure, h_c the convection coefficient, T_a the ambient air temperature and T_{wb} the temperature of the wet bulb. The heat lost by evaporation is equal to the heat gained by convection. There is no term for radiant heat exchange in equation (1), and the Assman or sling psychrometer reaches wet bulb temperature because the heat exchanges by evaporation and convection are made large in comparison with that by radiation.

For any wet bulb line on the psychrometric chart the wet bulb temperature is constant. The line is an isotherm showing a family of air temperatures and ambient water vapour pressures which correspond with the wet bulb temperature in question. By rearranging equation (1) the relation between the variables p_a and T_a for a given wet bulb line can be found.

$$p_a = p_{wb} + T_{wb} h_c/h_e - T_a h_c/h_e. \tag{2}$$

It was shown in Chapter 2 that the ratio h_c/h_e is independent of wind speed and size and shape of the surface. For a given wet bulb temperature the first two terms on the right-hand side of equation (2) are constant, so the relation between p_a and T_a is linear and independent of air movement. The slope of the line is $-h_c/h_e$ and it passes through the point $p_a = p_{wb}$, $T_a = T_{wb}$, where evaporative and convective heat exchanges are zero.

Unheated wet surface with radiation

The heat balance of a wet surface which is exchanging heat with the environment by evaporation, convection and radiation can be conveniently represented using the Operative temperature terminology (p. 66).

$$h_e(p_s - p_a) = h_o(T_o - T_s). \tag{3}$$

This equation describes the temperature, T_s, of a wet surface which is losing heat by evaporation at the same rate as it is gaining heat by convection and radiation. The Operative temperature, T_o, is a suitable combination of the air and radiant temperatures of the environment, and the Operative heat transfer coefficient, h_o, is the sum of the convection coefficient, h_c, and the first power radiation coefficient (including emissivity and area factors), h_r. By analogy with equation (1) it will be seen that on the psychrometric chart the isotherms for the wet surface will be parallel lines of slope

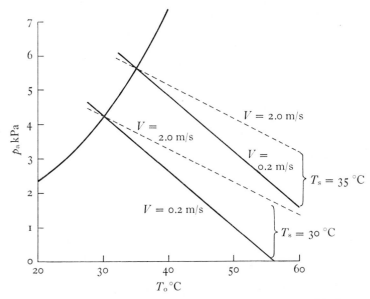

Fig. 4.2. Isotherms for an unheated wet surface exchanging heat with the environment by evaporation, convection and radiation. The Operative temperature, T_o, is a suitable combination of air temperature and mean radiant temperature. The slopes of the lines depend on the wind speed, approaching the slope of the wet bulb lines at high wind speeds.

$-h_o/h_e$. However one of the constituents of h_o, namely h_c, depends on air movement and is proportional to h_e at all wind speeds, while the other, h_r, is independent of wind speed. The ratio h_o/h_e therefore depends on wind speed and different lines are required for different wind speeds. (Where air temperature and mean radiant temperature differ, air movement also enters the expression for Operative temperature.) Examples for wind speeds of 0.2 and 2.0 m/s are shown in Fig. 4.2. The heat exchange coefficients used in constructing this figure are appropriate for a man standing in

4. The heat balance of sweating skin

a wind. The lines for $V = 2.0$ are less steep than those for the lower air movement and are closer to the wet bulb lines. All the lines for a given surface temperature meet at the point $T_0 = T_s$, $p_a = p_s$, where the evaporative heat exchange and the net sensible heat exchange are both zero. The line for saturation at T_0 has been drawn in Fig. 4.2, and if air and radiant temperatures are equal it represents saturation of the ambient air. If the walls are cooler than the air it is possible for the ambient water vapour pressure to be higher than saturation at T_0, since T_a must then be higher than T_0. The situation is uncommon in enclosed spaces, where condensation on the walls would occur, but may be encountered at night in the open, when the radiant temperature of the sky may be much lower than air temperature. Dew will then form on the surface, the heat liberated by condensation balancing the net loss of sensible heat. Both sides of equation (3) become negative.

Wet surface losing heat

In the cases so far considered there was no net heat exchange between the surface and the environment. This is an unusual state for human subjects since the body is producing heat and the physiological responses tend to promote heat balance with a net loss of heat to the environment. If this net heat loss from the skin is designated H_s, the equation for heat balance at the skin surface (not necessarily for heat balance in the body, since H_s may not equal the requirement for body heat balance) becomes

$$h_e(p_s - p_a) = h_0(T_0 - T_s) + H_s. \qquad (4)$$

The isotherm for a wet surface losing heat is parallel with that for a surface having no net heat transfer (slope $= -h_0/h_e$) but the line is displaced on the temperature axis by H_s/h_0, the temperature difference required to produce a sensible heat loss equal to H_s. The line can equally well be regarded as displaced on the vapour pressure axis by H_s/h_e, the vapour pressure difference required to produce an evaporative heat loss equal to H_s.

Examples of 35 °C isotherms are shown in Fig. 4.3 for air movements of 0.2 and 2.0 m/s and a net heat loss H_s, of 100 W/m². Isotherms for $H_s = 0$, identical with those in Fig. 4.2 are included for comparison.

If the Effective Radiant Field method is used to express the radiant heat load (p. 73), the temperature axis is entered with air temperature. The extra heat gained by radiation, H_r, must be lost in addition to the skin heat loss, H_s, so the displacement of the isotherms from those for $H_s = 0$, $T_0 = T_a$, is now

$$(H_s + H_r)/h_o \text{ }^\circ\text{C} \quad \text{or} \quad (H_s + H_r)/h_e \text{ kPa.}$$

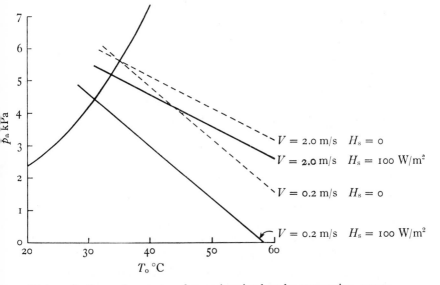

Fig. 4.3. Isotherms for a wet surface exchanging heat by evaporation, convection and radiation. All the lines refer to a surface temperature of 35 °C. The dashed lines are isotherms for a surface having no net heat exchange; the continuous lines for a surface losing heat at the rate of 100 W/m². Isotherms at each wind speed are parallel, but when heat is lost the line is shifted to the left.

The equilibrium temperature of wet skin

Equation (4), describing the heat balance of a wet surface, contains temperature and vapour pressure terms. If the skin temperature is known, its vapour pressure is also known and the equation can be solved for any of the other terms. However if the skin temperature is the unknown quantity the equation cannot be solved because skin temperature and skin vapour pressure are both variables. It is necessary to express one of these in terms of the other in order that the equation may contain only one unknown term. Empirical equations relating saturated water vapour pressure to temperature over

79

4. The heat balance of sweating skin

a wide range are cumbersome, but over a few degrees a linear approximation of the form $p = a.T + b$ can be used without introducing much error. The values of a and b depend on the temperature range. Some examples in the general range 30–40 °C are shown in Table 4.1.

TABLE 4.1. *Linear approximations for vapour pressure, p, in terms of temperature, T.*

Temperature range °C	Equation	Maximum error kPa
30–34	$p = 0.270(T-30) + 4.23$	0.02
34–38	$p = 0.325(T-30) + 4.00$	0.02
30–40	$p = 0.312(T-30) + 4.13$	0.12

When an expression of this form is substituted in equation (4), the equation for skin temperature becomes

$$T_s = \frac{h_0.T_0 + h_e(p_a - b) + H_s}{(h_0 + a.h_e)}. \qquad (5)$$

This can be used to calculate the mean skin temperature at equilibrium of a human subject whose skin is completely covered with a film of sweat (Machle & Hatch, 1947). When equilibrium is reached the total heat loss must be equal to the rate of metabolic heat production less the rate of external working. It must also equal the sum of the skin heat loss, H_s, and the heat loss from the respiratory tract (likely to be very small in conditions consistent with completely wet skin). A direct experimental check of equation (5) for the steady state is possible but technically difficult. The range of environments which could be covered is very restricted, because the stress must be severe enough to ensure that the skin is completely wet, and at the same time the subject must be able to tolerate the conditions long enough to reach thermal equilibrium.

The approach to equilibrium

By a simple rearrangement, equation (5) can be used to predict the rate of heat loss at any skin temperature. Metabolic heat which is not lost is stored in the body, and the rate of storage, S, is given by

$$S = (M - W) - H_s. \qquad (6)$$

Heat loss from the respiratory tract has been neglected, but could be included in the bracket. If heat is being stored, temperatures throughout the body will rise, and if a relation could be found between the rate of heat storage and the rate of rise of skin temperature, the way in which the latter approaches equilibrium could be predicted.

The stored heat is not necessarily distributed evenly throughout the body. Recognizing this, Machle & Hatch (1947) suggested that it might be justifiable to assume that the proportion stored in the superficial tissue remains constant as the temperatures rise. Thus the ratio of the change in skin temperature to the change in mean body temperature would be constant and could be assigned the symbol, n. The rate of change of skin temperature, $\dot{T}_\mathrm{s}$, would be then given by

$$\dot{T}_\mathrm{s} = n \cdot S/C \qquad (7)$$

where C is the heat capacity of the body per unit skin area. Equations (5), (6) and (7) can be combined to give an expression for $\dot{T}_\mathrm{s}$ in terms of T_s, M, W and the environmental conditions. This expression can be integrated to provide an equation for the skin temperature at any time after entering the environment when the skin is completely wet. It is simplest to introduce the final equilibrium skin temperature, T_s^*, given by equation (5).

$$(T_\mathrm{s}^* - T_\mathrm{s}) = (T_\mathrm{s}^* - T_\mathrm{s0}) \exp\left(-\frac{n}{C}(h_0 + a \cdot h_\mathrm{e})\, t\right). \qquad (8)$$

Here T_s0 is the skin temperature at zero time, again with the skin completely wet. Equilibrium is approached in a simple exponential fashion provided that n is constant. The term $(h_0 + a \cdot h_\mathrm{e})$ is the increase in heat loss by all three channels, radiation, convection and evaporation combined, per degree rise in skin temperature.

Experimental observations of the approach of skin temperature to its equilibrium value in different environments are shown in Fig. 4.4. In these experiments the conditions were such that mouth temperatures and skin temperatures rose at the same rate. It was therefore considered justifiable to assume a value of $n = 1.0$ for the distribution of stored heat. The slope of the lines on a semilogarithmic plot should then be $(h_0 + a \cdot h_\mathrm{e})/C$ and should be independent of air temperature and humidity. All the experiments were performed at the same air movement, so one would expect the lines to be parallel. They do not differ significantly in slope, and the mean slope provides an estimate for $(h_0 + a \cdot h_\mathrm{e})$ of 34.6 W/m². °C. The estimate derived from measurements of h_0 and h_e on a dummy in the same situation as the subject was 33.9 W/m². °C. This con-

4. The heat balance of sweating skin

firms that equation (8) is a satisfactory description of the approach to equilibrium under these conditions and that equation (5) correctly predicts the final equilibrium temperature.

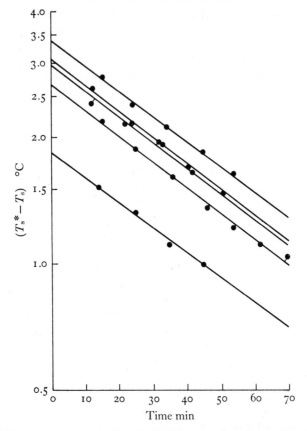

Fig. 4.4. Experimental observations of the approach of skin temperature to its equilibrium level. Each line refers to a different environment. All environments were hot and humid enough to produce complete wetting of the skin. Air movement was constant. The calculated equilibrium skin temperature, T_s^*, was found from equation (5) in the text. (From Kerslake & Waddell, 1958.)

Weighting of wet and dry bulb temperatures

The isotherms for wet surfaces are all straight lines on the psychrometric axes and can be defined by slope and position. They can also be defined approximately in terms of wet and dry bulb temperatures, and wet and dry bulb weighting factors are sometimes used, directly or indirectly, in the construction of indices of heat stress. It is therefore of some interest to examine the extent to which isotherms may be expressed in terms of wet and dry bulb weightings.

It will be seen from Fig. 4.1 that the wet bulb lines, although parallel, are not equally spaced on the psychrometric chart. At any dry bulb temperature the increment of ambient water vapour

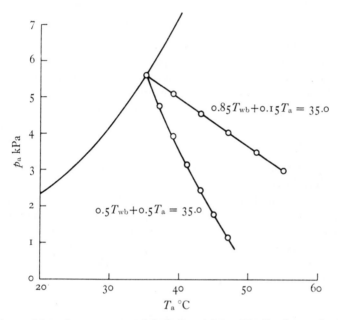

Fig. 4.5. Lines of constant wet and dry bulb weighting. The line for equal weighting is distinctly curvilinear, but in the case of the weighting $0.85T_{wb}+0.15T$ the departure from linearity is trivial.

pressure corresponding to a change of 1 degC in wet bulb temperature is greater at higher wet bulb temperatures. This is a consequence of the shape of the curve for saturated water vapour pressure. It follows that if dry and wet bulb temperatures are combined by a simple weighting formula, lines in which the weighted mean is constant will be curvilinear on the psychrometric axes. Examples are shown in Fig. 4.5.

The greater the weighting assigned to the wet bulb, the more nearly will the line approach the straight wet bulb line. In Fig. 4.5

4. *The heat balance of sweating skin*

a straight line has been drawn for the weighting $0.85 T_{wb} + 0.15 T_a$ but the line for the weighting $0.5 T_{wb} + 0.5 T_a$ is distinctly curvilinear.

In examining the relation between wet and dry bulb weightings and wet surface isotherms it is simplest to begin by considering the case $T_r = T_a = T_o$ (i.e. the surrounding surfaces are at air temperature). The heat balance equations for wet skin and for the wet bulb are restated

$$\text{Skin} \qquad h_e(p_s - p_a) = h_o(T_a - T_s) + H_s, \qquad (9)$$

$$\text{Wet bulb} \quad h_e'(p_{wb} - p_a) = h_c'(T_a - T_{wb}). \qquad (10)$$

The wet bulb thermometer is forcibly ventilated, but the wet bulb isotherm is independent of wind speed, so we can imagine a wet bulb having no radiant heat exchange and ventilated in such a way that h_e' for the wet bulb equals h_e for the skin. In this case h_c' must be equal to h_c for the skin, and equations (9) and (10) can be combined.

$$h_e(p_s - p_{wb}) = (h_o - h_c) T_a - h_o . T_s + h_c . T_{wb} + H_s. \qquad (11)$$

This equation still contains both temperature and vapour pressure terms, but the latter can be removed by using the linear approximation for saturated vapour pressure in terms of temperature (Table 4.1). However the range of temperature is greater than that involved in the skin temperature application made previously, and the constants a and b appropriate for p_{wb} may not also be appropriate for p_s. In principle, therefore, one should allow different constants for the two cases.

$$p_s = a . T_s + b, \quad p_{wb} = a' . T_{wb} + b'.$$

When these expressions are substituted in equation (11), the skin temperature is given by

$$T_s = \frac{(h_c + a' . h_e) T_{wb} + (h_o - h_c) T_a + H_s + h_e(b' - b)}{(h_o + a . h_e)}. \qquad (12)$$

If temperature and humidity are to be combined by weighting the wet and dry bulb temperatures, the sum of the weighting factors must be unity. This is true for equation (12) only if $a = a'$, and it is therefore necessary for the skin and wet bulb vapour pressures to be described by the same linear approximation. In this event, $b = b'$ and the last term becomes zero. The consequent error diminishes as wet bulb temperature approaches skin temperature, and is small in hot, humid environments.

Weighting of wet and dry bulb temperatures

Figure 4.6 shows weighting factors for wet and dry bulb temperatures calculated from equation (12), using the approximation for the range 30–40 °C in Table 4.1. They are also shown in Appendix 4. The heat exchange coefficients are appropriate for a nude man standing in a wind. If the wet and dry bulb temperatures are weighted in proportions appropriate for the air movement

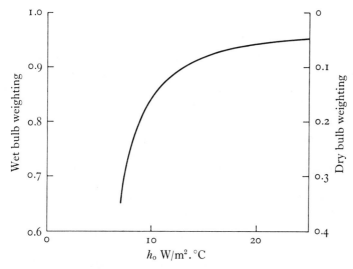

Fig. 4.6. Wet and dry bulb weightings appropriate for a wet surface. With no net heat exchange the temperature of the surface is equal to the weighted mean of the wet and dry bulb temperatures. The weighting depends on air movement and is here plotted against the Operative heat transfer coefficient.

the resulting mean is approximately equal to the skin temperature at which there would be no net heat exchange with the environment. Pairs of wet and dry bulb temperatures leading to the same weighted mean correspond to situations in which the skin temperature would be the same, i.e. they define the wet surface isotherms for a surface having no net heat exchange (Fig. 4.2). If the subject is losing heat his skin temperature will exceed the weighted mean of the wet and dry bulb temperatures by $H_s/(h_0 + a.h_e)$. If the mean radiant temperature is not equal to air temperature, H_r must be added to H_s.

Allowance for radiation can also be made using the Operative temperature approach, but this is complicated by the replacement

4. The heat balance of sweating skin

of T_a by T_o in equation (9). The difficulty can be overcome by introducing a term $(T_o - T_a)$ so that equation (9) becomes

$$h_e(p_s - p_a) = h_0(T_a - T_s) + h_0(T_o - T_a) + H_s. \qquad (13)$$

The final equation for skin temperature now contains both T_a and T_o. For the case $a = a'$, $b = b'$, it becomes

$$T_s = \frac{(h_c + a.h_e) T_{wb} - h_c.T_a + h_o.T_o + H_s}{(h_o + a.h_e)}. \qquad (14)$$

This provides weighting factors for wet bulb temperature, dry bulb temperature and Operative temperature (or globe temperature). It is noteworthy that the factor for dry bulb temperature is negative, and in this respect differs from the form of weighting known as the WBGT index (p. 238). The term $(h_o - h_c) T_a$ in equation (11), which relates to the condition $T_a = T_r$, can be expressed as $h_r.T_r$, as can the corresponding part of equation (14), $(h_o.T_o - h_c.T_a)$. These are terms describing radiant heat gain, and are added to H_s in the same way as Effective Radiant Field when this is used.

Representation on the psychrometric chart

The line represented by a weighting of wet and dry bulb temperatures,

$$n.T_{wb} + (1 - n) T_a = \bar{T},$$

must pass through the saturation line at $\bar{T}$, since at saturation the wet and dry bulb temperatures are equal. If it is considered acceptable to use a straight line to represent the equation (approximately) a single point elsewhere on the line is sufficient to define it.

The slope of the line can be found by working out two points and dividing the difference in vapour pressure by the difference in temperature, but it is sometimes simpler to use an algebraic formula which can be derived as follows.

The continuous straight line in Fig. 4.7 represents a weighting of wet and dry bulb temperatures such that $n.T_{wb} + (1 - n) T_a = \bar{T}$. The dashed lines are wet bulb isotherms. The point X, x is the saturated environment at $\bar{T}$, and the point Z, z some other condition in which the weighted mean of the temperatures is $\bar{T}$. The difference between the dry bulb temperatures in the two conditions, $z - x$, will be called ΔT_a, and the difference between the wet bulb temperatures, $x - y$, will be called ΔT_{wb}. These are related by the weighting equation

$$\Delta T_{wb} = \frac{(1 - n)}{n}.\Delta T_a.$$

The vapour pressure difference, $X - Y$, is the difference in saturated water vapour pressure between the two wet bulb temperatures, and is equal to $a.\Delta T_{wb}$, where a is the appropriate factor in Table 4.1. The vapour pressure difference, $Y - Z$, is $0.067 (\Delta T_{wb} + \Delta T_a)$, since the slope of the wet bulb lines is

-0.067 kPa/°C. The slope of the continuous line, $\Delta p_a/\Delta T_a$, is $(Z-X)/(z-x)$, and can be expressed in terms of n and a

$$-\Delta p_a/\Delta T_a = (a(1-n)+0.067)/n. \tag{15}$$

Similarly $$n = (a+0.067)/(a-\Delta p_a/\Delta T_a). \tag{16}$$

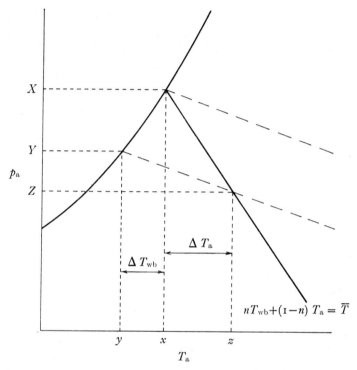

Fig. 4.7. Illustration of the method of calculating the slope of the line,

$$n.\,T_{wb}+(1-n)\,T_a = \overline{T},$$

on the psychrometric axes. The dashed lines are wet bulb isotherms. For explanation see text.

Sweating skin

So far the examination of the heat balance of sweating skin has been restricted to cases in which the skin is completely covered by a film of sweat. For this state to be maintained the liquid film must be replaced by sweat at least as fast as the water evaporates, and for a man it represents the case when at all points on the skin surface

4. The heat balance of sweating skin

the rate of sweat production exceeds the rate of evaporation. Woodcock, Pratt & Breckenridge (1952) examined the general case in which sweat is produced at a fixed rate which does not necessarily exceed the rate of evaporation.

Consider first a small region of skin with sweat rate S and local heat exchange coefficients h_o and h_e. The case $H_s = o$ (no net heat loss from the skin) is shown in Fig. 4.8. The point A represents the saturated environment ($T_a = T_r = T_o$) in which there will be

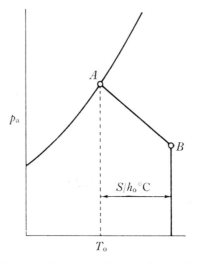

Fig. 4.8. Isotherm for a small region of skin sweating at a fixed rate and having no net heat exchange with the environment. The line AB is a wet surface isotherm for the appropriate surface temperature. Some of the sweat is wasted by dripping. At vapour pressures lower than point B all the sweat evaporates. The evaporative heat loss is then constant and equal to the sensible heat gain. Operative temperature is therefore constant for this part of the isotherm.

no heat loss. The skin is at ambient temperature so both sensible heat exchange and evaporation are zero. Since no sweat evaporates the skin will be completely wet. From this point the line for wet skin may be constructed as described earlier. The slope is $-h_o/h_e$ and at all points on the line the sensible heat gain, $h_o(T_o - T_s)$ equals the maximum evaporate heat loss, $h_e(p_s - p_a)$.

At some point on this line the rate of evaporation will be equal to the rate of sweat production, and at humidities lower than this the rate of sweat production, which is assumed to be constant, will be less than that required to keep the skin fully wet. In these circumstances the evaporative heat loss will be equal to the sweat rate, S (if changes in the heat of evaporation are ignored). This rate is independent of humidity, as is the rate of sensible heat exchange, so that this 'dry' portion of the curve is a vertical line defined by the heat balance $S = h_0(T_0 - T_s)$. The wet and dry lines meet at the point B where S is just equal to $h_e(p_s - p_a)$.

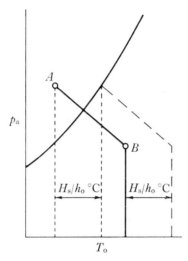

Fig. 4.9. Isotherm for a small region of skin sweating and losing heat at a constant rate, H_s (continuous line). The dashed line is the isotherm from Fig. 4.8 for the case in which there is no net loss of heat from the skin.

If there is a net loss of heat from the skin the isotherm is shifted to the left but is unaltered in other respects (Fig. 4.9) The position of the dry portion of the curve is now defined by

$$S = h_0(T_0 - T_s) + H_s,$$

so it is shifted to the left by H_s/h_0 °C. The wet portion is shifted by the same amount (p. 78). The point A is now at an Operative temperature H_s/h_0 °C below skin temperature and at a vapour pressure equal to saturation at skin temperature. The point may not be attainable in practice, but is useful for defining the position of the wet part of the isotherm. At this point there is no evaporative heat exchange and the sensible heat loss is equal to H_s.

4. *The heat balance of sweating skin*

So far only one sweat rate has been considered. If the sweat rate is greater the wet portion of the isotherm will be extended as shown in Fig. 4.10(*b*). The two examples shown in this figure are geometrically similar. The slopes of the wet and dry portions of the curves are the same in each case; it is only the point of intersection which differs, and in both cases this occurs at a temperature S/h_0 °C higher than point A and at a vapour pressure S/h_e kPa lower than point A.

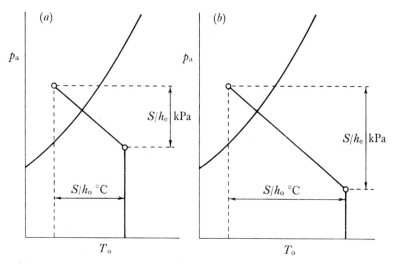

Fig. 4.10. The effect of a change in sweat rate on the skin isotherm. (*a*) is the same as Fig. 4.9. In (*b*) the sweat rate is greater. The wet part of the isotherm is longer and the Operative temperature at full evaporation is higher.

The form of these isotherms is generalized in Fig. 4.11. Here the skin temperature, sweat rate and environmental factors are combined into two ratios. W, the wettedness, is equal to E/E_{max}, where E is the rate of evaporation and E_{max} the maximum rate under the prevailing conditions (i.e. when the skin is completely wet). The other term, η_s, may be thought of as the efficiency of sweating, and is equal to E/S. If W is 1.0, η_s can have any value between o and 1.0; conversely when η_s is 1.0, W can have any value between o and 1.0. In the very simple context under consideration here this diagram may seem an unnecessary elaboration, but it provides a foundation for expressing the more complicated case of the whole human subject, which will now be examined.

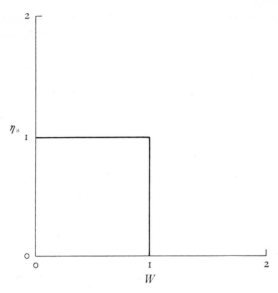

Fig. 4.11. Summary of the characteristics of the isotherms for small skin regions. The efficiency of sweating, η_s, is defined as E/S. W is the wettedness, E/E_{max}. If W is 1.0, η_s can have any value between 0 and 1.0; similarly if η_s is 1.0, W can have any value between 0 and 1.0.

Isotherms for the whole body

If it is assumed that skin temperature, sweat rate and all heat exchange coefficients are the same at all points on the body surface, isotherms for the whole body could be constructed which would have the form of those in Fig. 4.10 (Woodcock, Pratt & Breckenridge, 1952). This is not the case in practice, and between the situations of wet skin and full evaporation there is a transitional state in which the skin is wet in some places but not in others (Woodcock, Powers & Breckenridge, 1956; Woodcock & Breckenridge, 1965). Sibbons (1966) has called this the regime of restricted evaporation. (He also distinguishes a fourth regime in which there is no sweating.) It was shown in Chapter 2 that if sweat rates and heat exchange coefficients vary from place to place over the body one would expect the proportion of sweat evaporating, η_s, to be related to the wettedness, W. Fig. 4.11 shows this relation for one point on the skin surface, where, of course, the transition from full wetness to full evaporation is sudden. This line was a generalization of the form of the skin isotherms on the psychrometric axes,

91

4. *The heat balance of sweating skin*

the positions of which depend on sweat rate, skin heat loss and air movement. The corresponding relation for the whole body (e.g. that for a uniformly sweating cylinder shown in Fig. 4.13) is, by the same token, a generalization of the form of the mean skin isotherms for the whole body. A qualification about the distribution of skin temperature and of sweat production is necessary here, since a single curve relating η_s and W might not be appropriate for all mean skin temperatures and sweat rates.

Each isotherm refers to a certain skin temperature, sweat rate, skin heat loss and air movement. If values for these are assumed and a curve relating η_s to W is available (e.g. Fig. 4.11 or Fig. 4.13), the isotherms can be constructed either by tabulation or by a graphical method, which will be illustrated here for the case of the sweating cylinder.

Table 4.2 shows values of η_s for a cylinder at various values of W. Fig. 4.12 shows the stages in the construction of the 35 °C isotherm for a cylinder sweating uniformly at 180 W/m² and losing heat at 60 W/m². It is assumed to be in a transverse wind such that $\bar{h}_o = 15$ W/m². °C, $\bar{h}_e = 120$ W/m². kPa. (These values would not be consistent with the case of a human subject, whose radiant area is less than his total surface area.)

TABLE 4.2. *Values of η_s at various values of W for a uniformly sweating cylinder in a transverse wind.*

$W\ (= E/E_{\max})$	1.0	0.9	0.8	0.7	0.6	0.5	0.45
$\eta_s\ (= E/S)$	0.60	0.73	0.82	0.89	0.94	0.98	1.00

1. Plot point A at $p_a = p_s$; $T_o = T_s - H_s/\bar{h}_o$ in this case at $p_a = 5.62$ kPa, $T_o = 31.0$ °C.
2. Plot point B 10 degC higher than point A and 10 $\bar{h}_o/\bar{h}_e$ kPa lower, in this case at $T_o = 41.0$ °C, $p_a = 4.37$ kPa. Draw a straight line through these points. This is the wet skin isotherm.
3. Draw a vertical line at $T_o = T_s + (S - H_s)/\bar{h}_o$, in this case at 43.0 °C. This is the line for full evaporation, $\eta_s = 1.0$, $E = S$.
4. Plot points at the vapour pressure of point B (in this case 4.37 kPa) and 1, 2, 3, 4 and 5 degC lower. Draw lines through these points and point A. These lines correspond to wettedness of 0.9 to 0.5 respectively (Fig. 4.12(b)).
5. On each of these wettedness lines mark off the appropriate ratio from Table 4.2. On the wet skin line ($W = 1.0$) measure the distance from A to the point of intersection of this line with the full evaporation line and make a mark 0.60 of the way along. For the line $W = 0.9$ make the mark 0.73 of the way along, and so on, (Fig. 4.12(c)).
6. The isotherm corresponds with the wet skin isotherm as far as the point plotted on it in Fig. 4.12(c). It then passes through the other plotted points to join the line for full evaporation at values of W below 0.45.

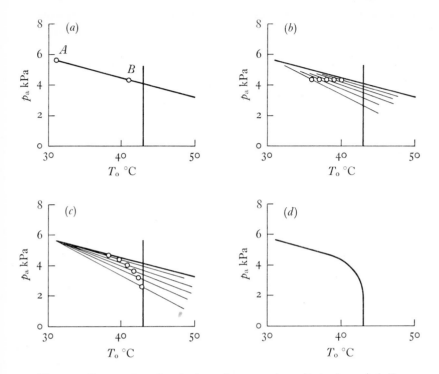

Fig. 4.12. Construction of an isotherm for a sweating cylinder in a wind. For details see text. At (a) the wet skin isotherm and line for full evaporation are constructed. Lines corresponding to different values of wettedness are contructed at (b), and one point on each is defined at (c). The final isotherm at (d) is drawn through these points and the appropriate parts of the wet skin and full evaporation lines.

Figure 4.13 shows the relations between η_s and W for a cylinder (continuous line) and for a single point on the surface (dashed line). The significance of these curves in relation to the mean skin isotherms can be appreciated by comparing Figs. 4.12(a) and (d). The former is the isotherm corresponding with the dashed line in Fig. 4.13, the latter that corresponding with the continuous line. Fig. 4.13 describes the way in which the corner of the isotherm is cut if the local heat exchange coefficients or sweat rates differ from place to place.

The isotherms developed so far have been based on the assumption that the sweat rate is constant when expressed in equivalent

4. *The heat balance of sweating skin*

heat units, W/m². Since the heat of evaporation increases as skin relative humidity falls this implies that the mass sweat rate is lower at lower skin relative humidities. The subject is therefore not in a constant physiological state. In making the transition from

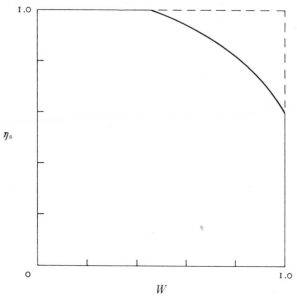

Fig. 4.13. Relations between η_s and W for a cylinder in a wind (continuous line) and at a single point on the surface (dashed line). This curve defines the shape of the appropriate isotherms, i.e. the form of the isotherm in the region intermediate between full wetness and full evaporation (cf. Fig. 4.12 (*a*) and (*d*)).

mean skin isotherms to lines of constant physiological state other factors must be considered too. The heat loss from the respiratory tract alters with ambient conditions as does the diffusion of water through the epidermis. These processes contribute to the subject's evaporative heat loss without making any demand on the sweating mechanism. They are considered in more detail in Chapter 8.

5 CLOTHING

Insulation

Clothing impedes the passage of sensible heat, both because of the insulation provided by the fabrics themselves and because of the layers of air trapped between the skin and clothing and between the various clothing layers. The insulation of fabrics is mainly due to the air contained in them, which constitutes a large proportion of their volume. The nature of the fibres does not exert much influence, and the insulation of many clothing materials can be estimated from their density (Peirce & Rees, 1945). This relation with density suggests that a significant amount of heat passes along the fibres, which conduct heat better than air. Hammel (1955) studied the conduction of heat through animal fur by replacing the air in the fur by Freon, which has different viscous and thermal properties. He concluded that conduction along the fibres was negligible and that the fur had less insulation than the same thickness of still air because convection currents were set up within it. When the hairs lay transverse to the heat flow the insulation was greater than when they lay parallel to it, and while this might suggest conduction along the hairs, he attributed the effect to differences in the orientation of the air spaces.

Thick fabrics designed for thermal insulation have a thermal resistance of about 0.25 °C.m²/W per *centimetre* thickness (conductivity 0.04 W/m.°C). The thermal resistance of air trapped between clothing layers is of the same order, and the total insulation of an assembly can be estimated by applying this figure to the total thickness, which can be found by measuring the girth of clothed and unclothed body segments (Siple & Cochran, 1944; Libet, 1945).

The mean insulation of a clothing assembly is sometimes expressed in 'clo' units (Gagge, Burton & Bazett, 1941). The clo is formally defined as a mean thermal resistance of 0.155 °C.m²/W. One clo will keep a resting man ($M = 58$ W/m²) comfortable in

95

5. Clothing

an environment at 21 °C, air movement 0.1 m/s, and can therefore be visualized as roughly the insulation of normal indoor clothing. This everyday yardstick can be misleading when multiplied. Few inexperienced people would suppose that the thickest arctic clothing assemblies have only about four times the insulation of indoor clothing. One tends to neglect the relatively large air spaces in light clothing.

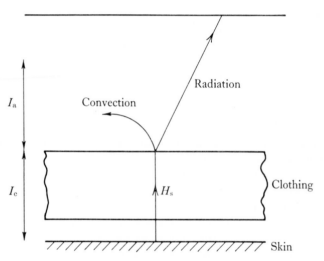

Fig. 5.1. Exchange of sensible heat between the skin and environment. The heat passes first through the clothing insulation, I_c, which includes the insulation of the garments and air spaces. From the clothing surface it passes to the environment by radiation and convection. These two channels are combined in I_a, the ambient insulation.

Sensible heat lost from the skin passes first through the insulation of the clothing assembly, I_c, and is then transferred to the environment by convection and radiation (Fig. 5.1). The heat transfer coefficients for convection and radiation at the clothing surface can be combined into a single figure, h_o, and expressed for convenience as I_a, the insulation of the environment. $I_a = 1/h_o$. The total thermal resistance between the skin and environment is then $I_c + I_a$. This is true at any point on the body surface, but if I_c and I_a vary from place to place it is not strictly correct to add their mean values in order to obtain the mean total insulation.

If the insulation of the clothing is assumed to be independent of wind speed and activity of the wearer, the total insulation at any speed can readily be found. Recommendations for the clothing required at different work rates and environmental conditions have been based on this (Belding, 1949; Burton & Edholm, 1955). In fact, the assumption that I_c is independent of wind speed is correct only at low air movements. In outdoor conditions, I_c changes with wind speed much more than does I_a (Breckenridge & Woodcock, 1950). Belding (1949) found that walking at a brisk pace halved the insulation of moderately thick clothing, probably because body movements pumped air in and out of the clothing.

Wind penetration

Air may enter clothing through apertures at the wrists, neck, etc. and thus circulate beneath the clothing (ventilation). It may also pass through the fabric or at least penetrate some distance into it, reducing its effective insulation. The latter process can be compared with that of rain falling on the surface of a pool. Heavy rain will cause turbulent disturbance in the upper layers of the water, but the deeper layers may remain still.

The penetration of wind into the coat of the horse has been examined by Tregear (1965). The wind flowed parallel with the surface of his sample and the hairs were aligned for minimum disturbance. Fig. 5.2 shows the temperature gradients through the coat in still air and at two wind speeds. The vertical dashed line indicates the surface of the coat, and points to the left indicate measurements made within its depth. At the higher wind speeds the temperature at the surface and a little below it is equal to the ambient air temperature. The wind must certainly have penetrated the most superficial part of the coat. The linear temperature gradients in the deeper layers suggest that the air there was not disturbed, and the depth of wind penetration is quite well demarcated.

Wind penetration of the coats of animals can markedly reduce their thermal resistance (Lentz & Hart, 1960; Bennett & Hutchinson, 1964; Bennett, 1964; Joyce, Blaxter & Park, 1966; Hutchinson & Brown, 1969). Wind striking the surface tends to part the hairs, producing regions of very low thermal resistance.

The coat of an animal grows out of its skin. Air cannot pass right through it. Moreover, since some hairs are longer than others,

5. Clothing

the density of the coat is greatest at the skin surface, and air does not readily flow in the deeper layers. Clothing is quite different in these respects. The layers least permeable to wind are usually on

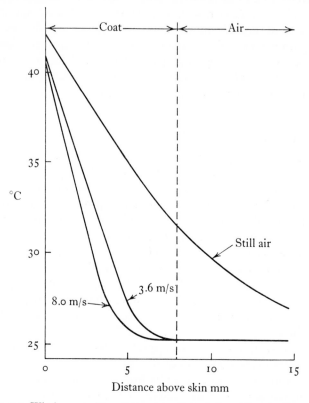

Fig. 5.2. Wind penetration of the coat of the horse. The abscissa represents distance through the coat and beyond, the skin surface being at the left-hand side. In still air the temperature gradient extends into the air outside the coat, but at moderate wind speeds the wind penetrates the outer part of the coat. (From Tregear, 1965.)

the outside, and there is often an air space between the inner surface of the clothing and the skin. Air which penetrates the outer layers is likely to pass right through the fabric to the skin surface where it can circulate freely. The loss of insulation in windy conditions depends on the wind permeability of the outer layer (Larose, 1947), but different assemblies having the same outer windbreak garment do not necessarily behave in the same way. Curves relating loss of insulation to wind speed differ in shape for different assemblies and it has not proved possible to describe them by means of a common equation (Breckenridge & Woodcock, 1950). It is usual to measure the total insulation, $(I_c + I_a)$, and to find I_c by subtracting the value that I_a would have if there were no wind penetration. However, if wind is known to penetrate the fabric, the concept of a boundary layer at the clothing surface can hardly be sustained. At wind speeds above about 2 m/s the changes in I_a are in any case trivial in comparison with those in I_c, even when clothing is designed for protection against wind.

Windbreaks

Fonseca, Breckenridge & Woodcock (1959) have examined the principles of windbreaks in fabric systems. Fig. 5.3 (a) shows a section through a cylindrical body around which are two cylinders of permeable fabric separated by air gaps. Wind which penetrates the outer fabric can either pass round the gap between the inner and outer fabrics or through the inner fabric into the space immediately surrounding the body. The partition will depend on the relative resistances to air flow of the two paths.

The overall heat transfer in a wind of 8 m/s was decreased if the inner space was filled with pile fibric (Fig. 5.3 (b)). Filling the outer space only (Fig. 5.3 (c)) increased the heat transfer about twofold. The resistance to passage of air round the outer space was increased, so more air penetrated the inner windbreak. When both spaces were filled (Fig. 5.3 (d)) the heat transfer coefficient was increased on the upstream side and decreased on the downstream side. It is of interest that the two windbreaks alone were so effective at reducing heat loss. The material used, Fortisan, is rather a poor windbreak, with a permeability about fourteen times that of Byrd cloth.

The variation in wind penetration characteristics of different clothing assemblies is readily comprehensible in the light of these findings, as is the efficacy of certain clothing combinations. Close-fitting underwear need not have an intrinsically low wind permeability, since it is backed by skin, and, like an animal's coat, cannot be completely penetrated by wind. Covering it by a loosely fitting trouser provides an assembly with a very high ratio of insulation to weight, minimal stiffness and good evaporative characteristics.

4-2

5. Clothing

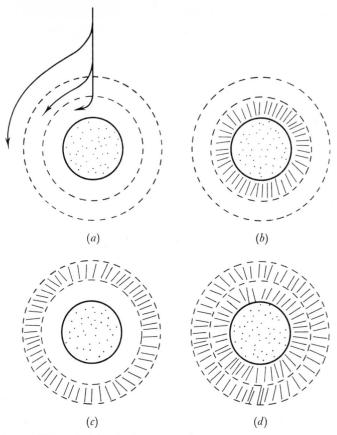

(a)   (b)

(c)   (d)

Fig. 5.3. The principle of windbreaks in clothing (Fonseca, Breckenridge & Woodcock, 1959). A heated cylinder is surrounded by two concentric cylinders of windbreak material. Diagram (a) shows how the wind is distributed through the system. At (b) the inner annulus is filled with pile fabric. This diminishes the heat loss from the cylinder. At (c) the outer annulus is filled. This increases the heat loss from the cylinder because more wind penetrates the inner windbreak. The total insulation at (d), when both annuli are filled, is about the same as at (a).

Dynamic insulation

If air is pumped into clothing through a distribution harness worn next the skin, it may escape by passing through the clothing fabric, and this increases the effective insulation of the clothing. The process has been called dynamic insulation. The simplest case, in which the air is liberated at skin temperature, has been examined theoretically by Spells (1960; 1961; 1966). The decrease in heat loss from the skin was found to depend on two dimensionless parameters, *m* and *n*. If *D* is the percentage decrease in skin heat loss (compared with the loss without ventilation),

$$1 - D/100 = (1+n) \cdot e^{-m}/(n+(1-e^{-m})/n). \qquad (1)$$

When the thickness of the clothing is small compared with the radius of the underlying body segment,

$$m = x.c.l/k, \quad n = k/h.l.$$

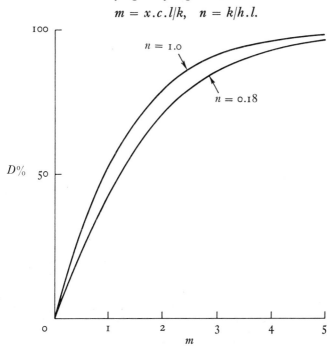

Fig. 5.4. Theoretical effect of air passing outwards radially through clothing. The percentage decrease in skin heat loss, *D*, is a function of two dimensionless parameters, *m* and *n* (see text). (From Spells, 1966.)

For thicker clothing both *m* and *n* require modification. In the above expressions, *x* is the mass flow of gas per unit area, *c* its heat capacity per unit mass, *l* the thickness of the clothing, *k* the thermal conductivity of the clothing with the ventilating gas as medium, and *h* the heat transfer coefficient at the outer surface of the clothing.

Equation (1) is shown graphically in Fig. 5.4. Values of *m* and *n* cover a realistic range, and it will be seen that the effect of *n* is not

5. Clothing

very great. For practical purposes it would be sufficient to express the dynamic insulation as a function of m only.

Spells & Blunt (1962; 1965) measured the dynamic insulation of various clothing assemblies applied to a cylindrical surface. Air and carbon dioxide were used for ventilation. The observed

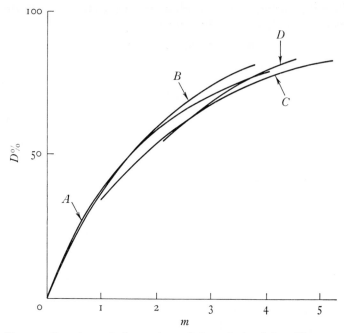

Fig. 5.5. Experimental observations of dynamic insulation. The percentage decrease in skin heat loss, D, is plotted against the parameter, m (see text).

    A. Air ventilation, clothing thickness 11 mm.
    B. Air ventilation, clothing thickness 21 mm.
    C. CO_2 ventilation; clothing thickness 11 mm.
    D. CO_2 ventilation, clothing thickness 21 mm.

(From Spells & Blunt, 1965; Spells, 1966.)

dynamic insulation effect (Fig. 5.5) was smaller than that predicted by theory, as would be expected if the air flow did not have the precise uniform radial character assumed in the theoretical analysis. However the general dependence of dynamic insulation on the parameter, m, was supported. The curves for the different clothing assemblies and ventilating gases are in good agreement.

Crockford & Goudge (1970) express the conductance of foamed

plastic material as an empirical function of thickness and air flow. They point out that at high temperatures, with large thermal gradients in the clothing, there may be significant heat transfer by radiation within the material, increasing the net conductivity.

Ventilation

In hot climates clothing may be necessary for reasons unconnected with thermal requirements, but it may serve as a valuable barrier to solar radiation and can reduce heat gain by convection. Except in special circumstances clothing intended for use in warm environments is designed to permit as much penetration by ambient air as possible, and by the use of loose closures exchange of air in the microclimate between the skin and clothing is encouraged. In such circumstances the nominal insulation of a clothing assembly as determined on a heated dummy at low air movement has little significance. Both external wind and bodily movement promote ventilation of the microclimate.

Figure 5.6 represents a clothing assembly in which there is ventilation between the skin and the inner clothing surface. It is assumed that radiation does not pass through the clothing. The skin loses heat by convection to the air in the microclimate and by radiation to the inner clothing surface. Some of the radiated heat is lost from the inner clothing by convection, the remainder passing through the clothing to the outer surface and thence to the environment by convection and radiation. If the ventilation beneath the clothing is large and the clothing thick, the inner clothing surface will be at air temperature. Radiation from the environment will not affect the skin heat exchange, and the clothing will act as a complete screen against radiation. If the radiant temperature of the environment is high and the air temperature is below skin temperature, increasing the ventilation will always diminish the heat stress. If the air temperature is above skin temperature and below the mean radiant temperature, there will be a certain ventilation rate at which sensible heat gain will be minimal. However, in such conditions evaporation will be an important factor and greater ventilation may be desirable.

The value of clothing as a radiation screen is demonstrated by the observations shown in Fig. 5.7. Clothing which is whiter than the skin will reflect more sunlight, thus reducing the radiation load,

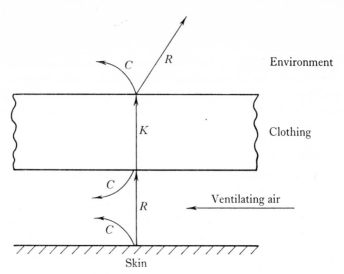

Fig. 5.6. Diagram of sensible heat exchanges in clothing. The space between the skin and the inner clothing surface is ventilated either naturally through the clothing fabric and closures or artificially.

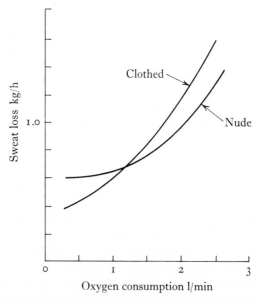

Fig. 5.7. The effect of clothing on the sweat loss of subjects in the desert. At the low work rates the clothing is beneficial because it acts as a radiation screen. At higher work rates this effect is overriden by restriction of evaporation. (From data of Robinson & Turrell, CMR Report No. 11 in Adolph in Newburgh, *Physiology of Heat Regulation and the Science of Clothing*. Philadelphia: W. B. Saunders Company, 1949.)

but this is unlikely to benefit the wearer unless there is also free circulation of air beneath the clothing. In the circumstances of the experiment the clothing was beneficial at low metabolic rates but became an embarrassment at higher rates, when the impediment to evaporation outweighed the benefit conferred by the radiation barrier.

When ventilation in the microclimate is fairly small and air spends an appreciable time in this layer, a further factor must be considered. Air enters the microclimate at the outside air temperature, but as it exchanges heat with the skin and clothing its temperature changes. The principles of this process are best illustrated by considering a simple case in which air is supposed to exchange heat only with the skin (and not with the inner clothing surface). The skin temperature is assumed to be uniform.

It is convenient to express the air temperature as the difference from skin temperature, and the symbol θ is used to emphasise that all temperatures are temperature differences. The abscissa represents the distance travelled by the air over the skin, and it is assumed that the cross section of the air space is constant, so that the air velocity is constant. The mass velocity of the air is $\dot{m}$. At any point, x, where the air temperature is θ_x, the air loses heat to the skin at a rate $h_c . \theta_x$. The temperature therefore falls as the air passes over the skin, and the curve in Fig. 5.8 is described by

$$d\theta/dx = -h_c . \theta_x / \dot{m} . c,$$

where c is the heat capacity of the air per unit mass. The expression for θ_x can be obtained by integration.

$$\theta_x = \theta_1 . \exp(-h_c . x / \dot{m} . c).$$

If the clothing and air flow are fixed, the exponential term will be constant. Calling this n for the value of x at the exit point, θ_2 will be $n . \theta_1$. The total heat exchange between air and skin is equal to the total heat taken up by the air, $\dot{m} . c(\theta_1 - \theta_2)$, which can also be expressed as $\dot{m} . c(1 - n) \theta_1$. Thus the heat exchange in this simplified example is the same as it would be if the air passed over the skin without suffering any change in temperature, and the convection coefficient were $\dot{m} . c(1 - n)$.

In the more complicated circumstances of real clothing assemblies the same principle can be applied, but the equations describing the heat exchanges become rather cumbersome (Kerslake, 1962;

5. Clothing

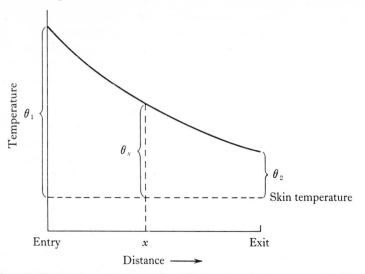

Fig. 5.8. Changes in temperature of ventilating air as it passes over the skin. At any point the air loses heat to the skin at a rate proportional to the temperature difference, θ, between the air and skin.

Burton, 1965). Moreover unless the air is actually pumped into the clothing (and preferably distributed there by means of an air ventilated suit) the air flow will not be known. Fortunately the general behaviour of an assembly, whether forcibly ventilated or not, can be empirically expressed very simply if the skin temperature is uniform (Kerslake, 1967 a). If the clothing surface is opaque to radiation,

$$H_s = k_0(T_s - T_0) + k_v(T_s - T_v). \tag{2}$$

Here k_0 and k_v are thermal conductances, both of which depend on the clothing assembly, external wind speed and ventilation. T_0 is the Operative temperature of the environment and T_v the temperature of the ventilating air at entry into the clothing assembly. Just as radiant temperature and air temperature can be combined in the Operative temperature, equation (2) can be rearranged to yield what may be called Clothed Operative temperature, T_{oc}, and heat exchange coefficient, h_{oc}.

$$H_s = h_{oc}(T_s - T_{oc}),$$
$$h_{oc} = k_0 + k_v,$$
$$T_{oc} = \frac{k_0}{h_{oc}} T_0 + \frac{k_v}{h_{oc}} T_v.$$

106

Ventilation

T_{oc} is a weighted mean of T_o and T_v, and T_o a weighted mean of $\bar{T}_r$ and T_a. In the case of natural ventilation, $T_v = T_a$. T_{oc} is then a weighted mean of $\bar{T}_r$ and T_a. The weightings differ from those for T_o, that for T_a being larger because ventilation provides a parallel channel for heat exchange with the air. The diminished weighting of $\bar{T}_r$ indicates the effect of clothing as a radiation screen.

In terms of the Effective Radiant Field (p. 73), $T_o = T_a + H_r/h_o$. For the case $T_v = T_a$, equation (2) becomes

$$H_s = (k_0 + k_v)(T_s - T_a) - \frac{k_0}{h_0} H_r. \tag{3}$$

The coefficient (k_0/h_0) represents the screening effect of the clothing, smallest when the screening is most effective. It can be thought of as an efficiency factor indicating the effect of H_r on H_s. It can never exceed unity because k_0, the effective conductance from the skin directly to the environment, can never be greater than h_0, the conductance from the clothing surface to the environment.

These conclusions are only true if radiation does not penetrate the clothing. This is a reasonable assumption if the outer layer is a closely woven windbreak, but in other assemblies radiation penetration may be significant. The network of thermal conductances is shown in Fig. 5.9. It is evident that if the conductances between skin and air are small the only channel for sensible heat exchange will be radiation, and T_{oc} will equal T_r. The rule that the weighting of T_r is reduced by clothing does not necessarily hold if the outer layer transmits radiation.

In sunlight the mean radiant temperature may be high and clothing can act as a radiation trap. The effect can be very large if the clothing is transparent to visible light but opaque to infra-red radiation. All the solar heat liberated at the skin surface must then be conducted away through the clothing, and it has been shown that T_{oc} could be some 50 degC higher than T_a (Burton & Edholm, 1955, p. 125). The necessary assembly is thick, so as to provide as much insulation as possible. The fashion houses commonly show a flagrant disregard of thermal requirements, but such transparent garments as they have produced are thin and, from a thermal point of view, quite safe.

Analysis of sensible heat transfer in clothing becomes very difficult if the skin temperature varies from place to place. The

5. Clothing

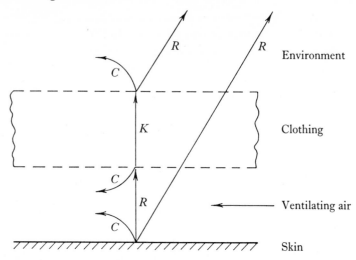

Fig. 5.9. Diagram of sensible heat exchanges when the clothing is not opaque to radiation. Radiant heat is exchanged with the environment from the skin surface, the clothing surface and from other levels within the clothing (not shown).

above equations are based on the assumption that it is uniform, and it is necessary to establish that realistic distributions of skin temperature do not invalidate them.

TABLE 5.1. *Values of k_0 and k_v at various ventilating air flows (see legend to Fig. 5.10). Both coefficients depend on the air flow.*

Ventilating air flow $\dot{m}.c$, W/m². °C	k_0 W/m². °C	k_v W/m². °C
0	1.70	—
3.6	1.35	2.96
5.9	1.25	4.65
8.4	1.12	6.20
11.3	1.01	7.95

Fig. 5.10 shows experimental results obtained on a heated dummy having sixteen independently controlled sections (Kerslake, 1964). Each section was arranged to reproduce the skin temperature/heat loss relation for the appropriate part of a comfortable resting subject (Clifford, 1966). The clothing assembly was forcibly ventilated at controlled temperature through an air

ventilated suit worn next to the skin. The constants k_0 and k_v were determined at various air flows and the linearity of the relations was confirmed (Fig. 5.10). Table 5.1 shows the values of k_0 and k_v. It will be noticed that k_0 is not independent of the ventilating air flow.

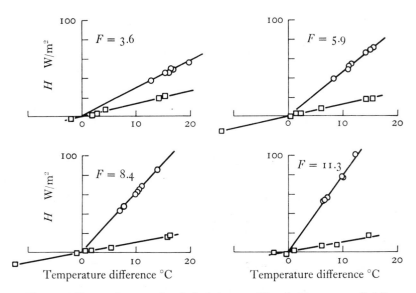

Fig. 5.10. Heat exchanges of a clothed dummy. The clothing was artificially ventilated with air at controlled temperature, T_v, and flow, F (expressed in terms of its heat capacity, W/m^2. $°C$). Skin heat loss was found to be proportional to $k_v(T_s - T_v) + k_0(T_s - T_0)$. The two components are plotted separately in each diagram, the abscissa being temperature difference. $\bigcirc$, $(T_s - T_v)$; $\square$, $(T_s - T_0)$. (From Kerslake, 1967a.)

Evaporation from wet skin

The principles used in the above examination of sensible heat transfer through clothing can also be applied to evaporation. Provided that the skin is completely wet and that the sweat evaporates directly from the skin and not from the clothing material vapour pressures can be substituted for temperatures and evaporation coefficients for thermal conductances in equation (2). If the clothing is ventilated by ambient air a single coefficient, the net vapour conductance between the skin and the environment, will suffice.

5. *Clothing*

Permeability index

Clothing materials impede evaporative heat transfer more than sensible heat transfer. If the conductivity of the material for water vapour (expressed in heat units) is called K_e, and that for sensible heat K_s, the ratio K_e/K_s expresses this property of the clothing.

Woodcock (1962) has developed a permeability index based on this ratio. His standard of comparison is air, for which the corresponding ratio is h_e/h_c. The permeability index of a material is $(K_e/K_s)/(h_e/h_c)$ and ranges from zero in the case of a material impermeable to water vapour, to unity in the case of air.

He has also applied the permeability index to the whole thermal enclosure. Denoting conductances for water vapour and sensible heat as k_e and k_s respectively, the permeability index is $(k_e/k_s)/(h_e/h_c)$. For the nude subject if radiation exchange is present the index is less than unity, being $(h_e/h_o)/(h_e/h_c)$. It would be attractive if the characteristics of the clothing could be separated from the environmental heat transfer coefficients, as for example by using (h_e/h_o) as the reference ratio. Unfortunately this will not work because k_e and k_s are neither independent of, nor proportional to h_e and h_c.

Permeation efficiency factor

Nishi & Gagge (1970) have proposed a different way of expressing the effect of clothing on evaporation. The permeation efficiency factor, f_{pcl}, is the ratio k_e/h_e, where k_e is the vapour conductance between the skin and air (expressed in heat units), and h_e that between the clothing surface and the air. The permeation efficiency factor depends on wind speed, clothing thickness and clothing material. The vapour permeability of cotton cloth was found to be almost equal to that of an equal thickness of still air, and it was suggested that the vapour permeability of normal clothing may be estimated from its thermal insulation. The authors stress that their analysis neglects the effects of moisture absorption by the clothing material, and that closely woven or vapour impermeable garments require special evaluation. They do not specifically refer to wind penetration or ventilation, both of which would affect the calculated relations between f_{pcl}, wind speed and clothing insulation.

Evaporation from incompletely wet skin

The analogy between sensible heat exchange and evaporation only holds if the skin is completely wet. The dashed curve in Fig. 5.11 shows how the vapour pressure of ventilating air changes as it passes over wet skin beneath clothing. (It is assumed that there is

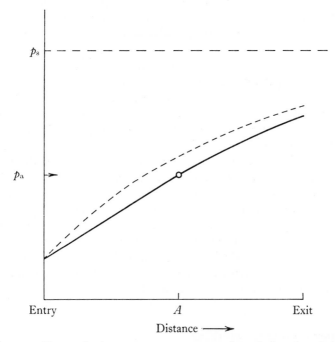

Fig. 5.11. Changes in the water vapour pressure of ventilating air as it passes over the skin. The dashed curve is for wet skin, when the rate of uptake of water vapour is proportional to $p_s - p_a$. The continuous curve is for incompletely wet skin. At first the rate of water vapour transfer is limited by the sweat rate. At A the humidity has risen to a level at which the local maximum evaporative capacity is equal to the sweat rate. Thereafter the curve is exponential, since the remainder of the skin is wet.

no vapour transfer through the clothing.) The curve has the same form as that of Fig. 5.8 for sensible heat exchange. In each case the rate of transfer depends on the local condition of the air.

However, where the local sweat rate is less than the local maximum evaporative capacity all the sweat evaporates, and the rate of evaporation is independent of the humidity of the air. This is

5. *Clothing*

illustrated by the continuous line in Fig. 5.11. In the first part of its passage the air takes up water vapour at a constant rate (equal to the sweat rate), and its vapour pressure increases linearly. At some point on the surface (A in Fig. 5.11) the vapour pressure may have risen to a level at which the local maximum evaporative capacity is equal to the sweat rate. The slope of the dashed curve represents the local maximum evaporative capacity, so the critical point, A, is reached at the vapour pressure indicated by the arrow, where the tangent to the dashed curve is parallel to the linear part of the continuous curve. Thereafter evaporation is limited by the humidity of the air, and the last part of the continuous curve follows the exponential curve for wet skin.

It was shown that the total sensible heat exchange for air passing through the microclimate can be expressed in terms of the air temperature at entry and an effective heat exchange coefficient which depends on the air flow. A similar effective evaporation coefficient, h'_e, can be used in the case of wet skin, but if the skin is not completely wet the value of h'_e depends on the sweat rate. If the sweat rate is very small the vapour pressure of the air will not change much as it passes over the skin and no part of the skin may become wet. h'_e will then be independent of the air flow and equal to $S/(p_s - p_a)$. The greater the sweat rate, up to that required for full wetting of the skin, the greater will be the effective evaporation coefficient. A term similar to the wettedness could be introduced to describe this, but is not helpful in practice.

Evaporation from clothing

It has so far been assumed that sweat evaporates at the skin surface and passes through the clothing in the form of water vapour. In practice liquid sweat may be transferred from the skin to the clothing and can then evaporate either from within the thickness of the clothing or from its surface. In either case the total rate of evaporation required for thermal balance is increased.

Consider a hypothetical case in which all the sweat evaporates from the same level in the clothing (Fig. 5.12). The temperature at the level where evaporation takes place is T_x. E_x is the rate of evaporation.

$$E_x = H_s - k_1(T_x - T_0), \quad H_s = k_2(T_s - T_x).$$

Combining these equations to eliminate T_x, and introducing k_0, the net conductance between skin and environment

$$H_s = k_0(T_s - T_0) + \frac{k_0}{k_1} E_x. \tag{4}$$

For evaporation from the skin, $k_1 = k_0$, so

$$H_s = k_0(T_s - T_0) + E_s. \tag{5}$$

Since the ratio (k_0/k_1) cannot exceed unity, evaporation from within the clothing must always be less effective, in terms of heat removed from the skin, than evaporation from the skin surface.

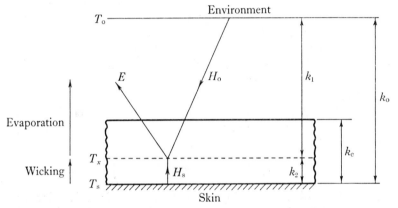

Fig. 5.12. Diagram of evaporation from within the thickness of clothing.

Efficiency of evaporation

In given conditions the amount of sweat required for heat balance is least when the sweat evaporates directly from the skin. This serves as a yardstick for expressing the efficiency of sweating when evaporation takes place within the clothing. From equation (4) the efficiency η_{sc} is equal to k_0/k_1. Givoni & Belding (1962) have suggested that η_{sc} is a function of E/E_{max}, as it is for the nude subject, when inefficiency is due to loss by dripping. They define E_{max} as the maximum rate of evaporation from wet clothing under the prevailing conditions (here called $E_{max.c}$), and present experimental evidence on one clothing assembly consistent with this proposition. Givoni & Berner-Nir (1967a) go further and propose that the same relation between $E/E_{max.c}$ and η_{sc} holds for all clothing assemblies and environments.

5. Clothing

The level at which evaporation takes place must be such that the maximum evaporative capacity at this level is just equal to the sweat rate. If the water table is higher than this, evaporation will exceed the rate of replacement and the level will fall. If k_{e1} is the vapour conductance superficial to this level,

$$E = k_{e1}(p_x - p_a).$$

Here p_x is the saturated vapour pressure at T_x (Fig. 5.12). Similarly, if we neglect changes in p_x with the level of evaporation,

$$E_{\text{max.c}} = h_e(p_x - p_a), \quad E/E_{\text{max.c}} = k_{e1}/h_e.$$

$E/E_{\text{max.c}}$ is independent of the clothing deep to the level of evaporation. η_{sc} does depend on this, and so cannot be defined solely as a function of $E/E_{\text{max.c}}$.

It remains possible that the proposed relation might hold for a given clothing assembly. It is simplest to assume that the ratio of vapour conductance to thermal conductance is the same in the clothing as it is in the boundary air layer (Nishi & Gagge, 1970). This allows $E/E_{\text{max.c}}$ to be expressed in terms of sensible heat transfer coefficients, and η_{sc} can be substituted for some of them. It can be shown that

$$E_{\text{max.c}}/E = 1 - \frac{h_c}{h_o} + \eta_{sc}\left(\frac{h_c}{k_c} + \frac{h_c}{h_o}\right).$$

If wind speed and clothing are constant, η_{sc} is a function of $E/E_{\text{max.c}}$, but this function changes if the wind speed changes and also depends on the clothing. If radiant heat exchange is ignored ($h_o = h_c$)

$$E_{\text{max.c}}/E = \eta_{sc}\left(1 + \frac{h_c}{k_c}\right).$$

This is simpler, but the dependence on wind speed and clothing remains.

If instead of taking the maximum evaporative capacity of wet clothing we use instead the maximum evaporative capacity at the skin surface, $E_{\text{max.s}}$, (which will be less than E if sweat is evaporating from within the clothing),

$$E_{\text{max.s}}/E = 1 - (1 - \eta_{sc})\frac{(1/k_c + 1/h_o)}{(1/k_c + 1/h_c)}.$$

If there were no radiant heat exchange, $E_{\text{max.s}}/E$ would be equal to η_{sc}. In practice h_o is not equal to h_c, so the relation between $E_{\text{max.s}}$ and η_{sc} depends on wind speed. However this relation is less dependent on clothing and wind speed than that using $E_{max.c}$. It is also more logical, since it contains a term concerned with the criterion for evaporation within the clothing, which will occur whenever E exceeds $E_{max.s}$.

Changes in vapour pressure at the level of evaporation have been neglected. This is necessary if vapour pressure terms are to be eliminated, but at high humidities the vapour pressure difference $(p_x - p_a)$ may be very sensitive to changes in p_x so the above equations are not very realistic. However, they serve to show that for clothed subjects the efficiency of sweating is unlikely to be a constant function of any term of the E/E_{max} type.

When only a single layer of thin clothing is worn, the position is somewhat simpler. If one's shirt is wet it is wet on both sides. Evaporation from the clothing takes place at a constant level, and is maintained by intermittent contact of the clothing with the skin, mopping up unevaporated sweat. For these conditions the rate of evaporation from the skin is $E_{max.s}$, and its efficiency is unity. The remainder of the sweat $(E - E_{max.s})$ evaporates from the clothing with efficiency k_0/h_0.

$$\eta_{sc} = E_{max.s}/E + \frac{k_0}{h_0}(1 - E_{max.s}/E).$$

This is a gross oversimplification, because $E_{max.s}$ is reduced if the clothing is wet. Also ventilation beneath the clothing has been neglected, and in a thin assembly this is likely to be an important factor. One would not expect these complications to remove the effects of clothing and wind speed on the relation between η_{sc} and $E_{max.s}/E$.

Regional differences in clothing, sweat production and heat exchange coefficients have not been considered, but it seems probable that in practice η_{sc} will be more closely related to $E/E_{max.s}$ than to $E/E_{max.c}$. Discrepancies might be reduced by including $E_{max.s}/E_{max.c}$ as another parameter (in the absence of radiation exchange this would be proportional to k_0/h_0). Measurement of these quantities in the laboratory is difficult, and it is unlikely that elaborate formulation would be justified. For field application and in the construction of heat stress indices the thermal properties of clothing can only be expressed very roughly, and the use of $E/E_{max.c}$ by Givoni & Berner-Nir (1967a) appears to work quite well.

6 RESPIRATION AND INSENSIBLE WATER LOSS

Respiratory heat exchange

Inspired air enters the respiratory passages at ambient temperature and humidity. The air leaving the alveoli in expiration is at the temperature of the pulmonary blood and saturated with water vapour, but to reach the exterior it has to pass through the bronchi, pharynx, and nose or mouth. The upper respiratory passages act as a regenerator of both sensible and latent heat. Inspired air, if cooler and drier than alveolar air, cools and dries these surfaces, which in turn cool and dry the expired air on its way out (Christie & Loomis, 1932; Seeley, 1940; Burch, 1945; Webb, 1951; Jackson & Schmidt-Nielsen, 1964). The process depends on the heat transfer coefficients between the air and the walls and on the effective capacities of these walls for heat and moisture.

Regenerators of sensible heat are used in engineering, where they sometimes have advantages over recuperators (heat exchangers which transfer heat from one fluid stream to another across a partition). The theory is well developed (Jakob, 1957), but cannot be applied to the respiratory passages where both sensible and latent heat are exchanged, the shape is complicated and the velocities constantly changing. The quantities of heat involved are not large in warm environments, although they may be important in the cold (Cole, 1954; Webb, 1955; Brebbia, Goldman & Buskirk, 1957; Walker & Wells, 1961), and the empirical studies of McCutchan & Taylor (1950) and of Aikas & Piironen (1963) provide an adequate basis for the practical calculation of respiratory heat and water loss in warm environments.

The total respiratory heat exchange can be expressed as the difference in enthalpy (total heat content per unit mass of dry air, J/g) between expired and inspired air. In both investigations this was found to correlate with the temperature and humidity of the inspired air. McCutchan & Taylor (1950) worked at ordinary room temperature, varying only the temperature and humidity of the inspired gas, whereas Aikas & Piironen (1963) used a hot chamber in which the subjects breathed ambient air. At temperatures up to 60 °C the results are in close agreement, and the discrepancies at higher temperatures may be related to the difference in the state of the subjects. Both sets of results are consistent with earlier work by Pfliederer & Less (1935) over a smaller temperature range.

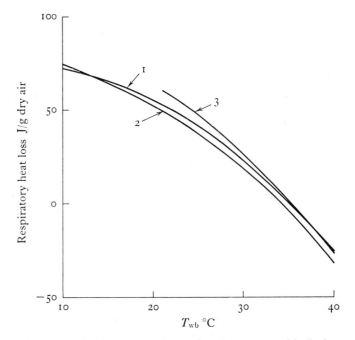

Fig. 6.1. Relation between respiratory heat loss, expressed in Joules per gram of dry air inspired, and the wet bulb temperature of the inspired air. 1, McCutchan & Taylor, 1950; 2, Aikas & Piironen, 1963, assuming saturated air; 3, Aikas & Piironen, 1963, assuming inspired air temperature 55 °C.

McCutchan & Taylor (1950) found that their results could be adequately described in terms of the ambient wet bulb temperature only (Fig. 6.1).

$$h_e - h_i = 73.8 + 0.682 T_{wb} - 0.0790 T_{wb}^2 \quad J/g. \tag{1}$$

Here h_e and h_i are the enthalpies of expired and inspired air (J/g) and T_{wb} is the ambient wet bulb temperature in °C. This is an empirical equation covering the range of wet bulb temperature

6. *Respiration and insensible water loss*

10–40 °C used in the experiments. Over this range the equation predicts that respiratory heat loss will decrease with increasing wet bulb temperature. The equation solves for $(h_e - h_i) = 0$ at $T_{wb} = 35.2$ °C, rather lower than one might expect for the case of saturated air unless blood going to the nasopharynx is considerably cooled by blood returning from the skin of the head. Saturated air was not used in the experiments, so it is not strictly justifiable to apply the equation to this condition.

Aikas & Piironen (1963) expressed the enthalpy change in terms of ambient temperature and vapour pressure, and a single line based on wet bulb temperature does not describe their results adequately. Fig. 6.1 shows curves calculated from their equation for saturated air and for a constant dry bulb temperature of 55 °C. Again it is unjustifiable to extrapolate their results in this way for saturated air, but the difference between the two lines shows that wet bulb temperature remains a fairly good guide to respiratory heat loss and that there is substantial agreement between the two investigations over the range shown.

If the volume of inspired air is measured, it is convenient to express the respiratory heat exchange in terms of the volume of inspired air at ambient temperature and pressure. The mass of dry air per litre inspired air depends on the temperature and humidity and is not defined by the wet bulb temperature. Fig. 6.2 is based on the equation given by Aikas & Piironen (1963).

$$(h_e - h_i) = 98.7 - 0.658T_a - 14.1p_a \text{ J/g}. \tag{2}$$

When the respiratory minute volume is not measured it can be estimated roughly from the rate of working. Subjects show variations of some 20 per cent about the mean, which is approximately given by $\dot{V} = 24\dot{V}_{O_2}$ where the minute volume, $\dot{V}$, is expressed at body temperature and ambient pressure saturated with water vapour. (The relation is almost independent of pressure (Ernsting 1965).) The oxygen consumption is closely related to the rate of working and the metabolic heat production. For external work at 20 per cent efficiency the heat production is about $270\dot{V}_{O_2}$ watts. The error of assuming this simple proportionality is small compared with the individual variations in minute volume, and if this level of accuracy is accepted one can express the minute volume in terms of metabolic heat production

$$\dot{V} = 0.09A_D(M - W) \text{ l/min}. \tag{3}$$

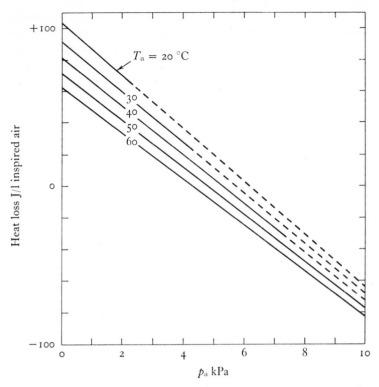

Fig. 6.2. Relation between respiratory heat loss, expressed in Joules per litre inspired air, and ambient water vapour pressure, p_a. Lines are shown for dry bulb temperatures, T_a, of 20 to 60 °C. The dashed parts of the lines refer to unreal situations in which p_a is greater than saturation at T_a. The diagram is based on the work of Aikas & Piironen (1963), equation (2) in the text.

Since M and W are conventionally expressed per unit skin area, the total heat production is $A_D(M - W)$. The minute volume so specified is at body temperature and ambient pressure, so that if the pressure is fixed the mass respiratory heat exchange is also fixed. (The accuracy of the estimate is insufficient to justify adjustment for changes in body temperature.) The respiratory heat loss can therefore be expressed as a proportion of the total heat production (it will be the same proportion if heat production is expressed per unit area) and this proportion will depend on the ambient wet bulb temperature (McCutchan & Taylor, 1950). The relation is shown for standard atmospheric pressure in Fig. 6.3. Individual variations

6. *Respiration and insensible water loss*

in minute volume at given work rate are about ± 20 per cent, so the prediction of respiratory heat loss is subject to this error. As, however, the respiratory heat loss is not more than about 10 per cent of the total heat production (Fig. 6.3), the proportion is predictable within about ± 2 per cent of the total heat production.

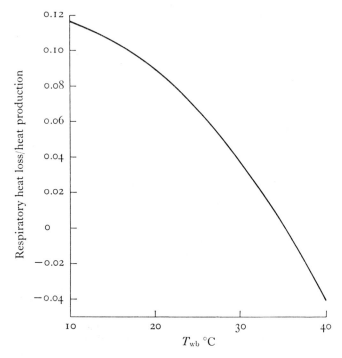

Fig. 6.3. Relation between respiratory heat loss, expressed as a proportion of body heat production, and wet bulb temperature of the inspired air.

Respiratory water loss

In considering the heat balance of the subject, nothing is to be gained by partitioning the respiratory heat exchange into its sensible and latent heat components. However the measurement of skin water loss is commonly based on observation of changes in the subject's weight, from which must be subtracted the loss of mass through respiration. The respiratory water loss is an important component of this.

The quantity of water taken up by the air breathed was expressed by McCutchan & Taylor (1950)

$$m_w = 0.0276 + 0.0000650T_a - 0.798w_a \text{ kg/kg dry air.} \quad (4)$$

Here the humidity of the inspired air is expressed as the humidity ratio, w_a, (see p. 11). The temperature term is not negligible, but if the water loss is expressed in terms of the volume of inspired air (rather than the mass), and the humidity as vapour pressure, the temperature term becomes very small. Over the range $T_a = 20 - 60 \,^\circ\text{C}$, $p_a = 0-5.0$ kPa, the water loss is nearly proportional to the ambient water vapour pressure (within 2 per cent).

$$m_w = 0.034 - 0.006p_a \text{ g/l inspired air} \\ \dot{m}_w = (2.04 - 0.36p_a)\,\dot{V} \text{ g/h.} \qquad \left.\right\} \quad (5)$$

At $\dot{V} = 50$ l/min this equation leads to a respiratory water loss of 102 g/h in dry air. This figure is unlikely to be exceeded, and the 2 per cent error in equation (5) leads to an error of not more than 2 g/h in the estimate of the respiratory water loss. If the minute volume is not measured, but estimated from the work rate, the error in the estimate of respiratory water loss can be 20 per cent because of variations between subjects. However, provided that the sweat rate is not less than the rate of heat production (i.e. at ambient temperatures above 35 °C) the error, even in dry air, will not exceed 3 per cent of the rate of weight loss.

Metabolic weight loss

In the steady state there is no overall exchange of nitrogen, but the mass of carbon dioxide lost is greater than the mass of oxygen consumed. If the oxygen consumption, $\dot{V}_{O_2}$, is expressed at standard temperature and pressure, the mass rate of oxygen consumption is $(32/22.4)\dot{V}_{O_2}$ g/min. At a respiratory quotient of 0.82, the mass of carbon dioxide expired is $0.82\,(44/22.4)\dot{V}_{O_2}$ g/min. The difference between these rates of mass transfer is the rate of metabolic weight loss, $\dot{m}_m$. $\quad \dot{m}_m = 0.18\dot{V}_{O_2}$ g/min or $11\dot{V}_{O_2}$ g/h. $\quad (6)$

An error of 10 per cent in the estimate of $\dot{V}_{O_2}$ or respiratory quotient would lead to an error of only about 2 g/h at an oxygen consumption of 2.0 l/min. For the purpose of calculating the metabolic weight loss a very rough estimate of oxygen consumption will suffice.

6. *Respiration and insensible water loss*

Cutaneous gas exchange

A small proportion (probably less than 2 per cent) of the metabolic gas exchange takes place through the skin (Fitzgerald, 1957). The metabolic weight loss through the skin is negligible, but water can pass through the skin in appreciable quantities even in the absence of sweating.

Insensible perspiration

This term is used here to mean loss of water from the skin which is not due to sweat gland activity (Robinson, 1949). The process is closely akin to a diffusion of water from tissue fluid at skin temperature through the epidermis to the ambient air (Trolle, 1937; Burch & Winsor, 1944; Taylor & Buettner, 1953; Heerd & Ohara, 1962), but the question of whether any active barrier or pump is involved has been the cause of some controversy (Lobitz & Daniels, 1961). Below the stratum corneum the epidermis is a wet tissue, and the main barrier to evaporation is probably in the lower layers of the stratum corneum (Monash, 1957; Monash & Blank, 1958), although all layers contribute to the vapour resistance (Potter, 1966). The stratum corneum is able to absorb large quantities of water (Blank, 1952; 1953) and one might expect the permeability to depend on the water content (King, 1945; Goodman & Wolf, 1969). If so, the relation between water loss and vapour pressure difference in the absence of an active barrier would be curvilinear, and one could not make deductions about the presence or absence of such a barrier from observations made at large vapour pressure differences (Buettner, 1959b). To detect the barrier it is necessary to make measurements of water transfer at ambient vapour pressures close to saturation at skin temperature. If the whole subject were exposed to such conditions he would normally sweat, and the diffusion component of his water loss could not be measured. This can be avoided by examining small regions of skin with a capsule technique (Pinson, 1942).

Buettner (1953) found an apparent gain of water by the skin when the vapour pressure in the capsule air exceeded 3.5 kPa. The vapour pressure of the tissue fluids was much higher than this and some form of active process appeared to be involved. However in later work (Buettner, 1959a, b) he obtained water losses from the skin at vapour pressures as high as 4.7 kPa. Dirnhuber & Tregear

(1960) used an ingenious device whereby the vapour pressure in the capsule could be varied and the water exchange measured without changing the air or removing the capsule. Their results were similar, water being apparently taken up by the skin from saturated air at skin temperature. The vapour pressure at which there was no water exchange was about 0.5 kPa below saturation at skin temperature, well below the vapour pressure of the tissue fluid. Tregear (1966, pp. 46–8) has pointed out that the apparent rate of inward transport in their experiments (0.6–1.0 μg/cm^2.min) might be accounted for by sideways transfer at the edge of the capsule, and that the quantity is small compared with the mass of water taken up by the stratum corneum (5000 μg/cm^2 at 90 per cent relative humidity). It was impossible to be sure that the water content of the stratum corneum had reached equilibrium, and he concluded that the evidence for an active water transport system in the epidermis was inconclusive.

There are several parallel routes whereby water might pass through the skin; through the epidermis, the hair follicles or the sweat ducts. These are discussed by Mali (1956) who concludes that direct passage through the epidermis may not be the most important process. Observations on subjects with congenital absence of sweat glands show that the rate of insensible water loss is similar to that of normal subjects who are not sweating (Loewy & Wechselmann, 1911; Richardson, 1926; Sunderman, 1941; Felsher, 1944). Normal subjects in a thermally neutral environment do not sweat from the greater part of the skin surface, but the glands on certain regions (palm, sole, axilla, perineum) are constantly active (non-thermal sweating). The total skin water loss includes the sweat produced by these glands.

Measurements of total skin water loss from resting subjects in the absence of thermal sweating over the skin temperature range 31–35 °C by Zöllner, Thauer & Kaufmann (1955) and by Brebner, Kerslake & Waddell (1956) are shown in Fig. 6.4. The latter group attempted to prevent evaporation of non-thermal sweat, but the similarity between the two sets of results suggests that they were unsuccessful. The straight line in the figure is within ± 3.0 g/m^2.h of all the points. It is described by

$$\dot{m}_{\mathrm{is}} = 6.0 + 1.75(p_{\mathrm{s}} - p_{\mathrm{a}}) \ \mathrm{g/m^2.h.} \tag{7}$$

This can readily be interpreted as a passive diffusion process

6. *Respiration and insensible water loss*

coupled with a steady sweat rate of 6.0 g/m^2.h. In terms of evaporative heat loss, neglecting changes in heat of evaporation, equation (7) can be expressed:

$$E_{is} = 4.0 + 1.2(p_s - p_a) \text{ W/m}^2. \qquad (8)$$

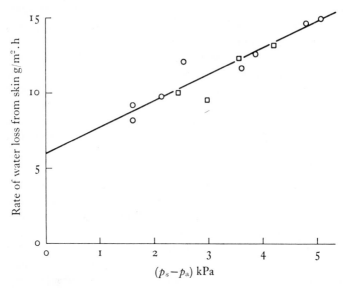

Fig. 6.4. Rate of water loss from the skin of nude subjects in the absence of thermal sweating. O, Zöllner *et al.* 1955; □, Brebner *et al.* 1956. The abscissa shows the difference between saturated water vapour pressure at skin temperature, p_s, and ambient water vapour pressure, p_a.

Webb, Garlington & Schwarz (1957) used massive doses of atropine to prevent sweating, and thus extended their observations to skin temperatures of 36.9 °C. The ambient air was nearly dry in all their experiments, and the vapour pressure difference between skin and air was confounded with skin temperature. Results at skin temperatures of 32.2 and 33.9 °C were not very different from those in Fig. 6.4 but at 36.9 °C (vapour pressure difference 5.5 kPa) the skin water loss was 32 g/m^2.h. Careful observation did not disclose any sweat gland activity and one might expect the results to be lower than those in Fig. 6.4 because of inhibition of non-thermal sweating. At the lowest skin temperature of 29.1 °C (vapour pressure difference 3.5 kPa) their subjects lost an average of about 5 g/m^2.h consistent with the line in Fig.6.4

if non-thermal sweating is assumed to be zero. Zöllner *et al.* (1955) found that the skin water loss at given vapour pressure difference was lower at lower skin temperatures, and interpreted this as a change in the permeability of the epidermis. One could also suppose that when the skin is cool, non-thermal sweating is diminished, possibly by a direct effect of temperature on the glands (p. 142).

The coefficient, 1.2 W/m^2.kPa in equation (8) is small compared with the evaporative coefficient at the skin surface (Appendix 4). The resistance of the epidermis to the passage of water vapour is so much greater than that of the air above it that one would not expect wind speed to have much effect on insensible water loss. DuBois (1936) reported that the rate of weight loss increased from 30 to 37 g/hr when air movement was increased by means of a fan, but this was associated with an increase in total heat loss. Evaporative heat loss was actually a smaller proportion of the total heat loss when the fan was used.

Equations (7) and (8) describe the skin water loss in the absence of thermal sweating. When the sweat glands are active, the vapour pressure at the skin surface is increased. The vapour pressure difference between the tissues and the skin surface is now $p_s - \phi_s \cdot p_s$. Since the epidermis provides nearly all the resistance to diffusional water loss, the coefficient, 1.2 W/m^2.kPa, in equation (8) can be retained, and the water loss, expressed in equivalent heat units, is given by

$$E_{is} = 1.2(1 - \phi_s) p_s \quad W/m^2. \tag{9}$$

Total insensible water loss

At ordinary room temperatures the total insensible water loss (from skin and lungs) is a fairly constant proportion of the metabolic rate. DuBois (1924) gives a figure of 23–27 per cent, and there is general agreement that about a quarter of the metabolic heat is lost in this way (Levine & Marples, 1940; Newburgh *et al.* 1937). The preceding discussion would lead one to expect that ambient humidity would have an appreciable effect and that the proportion would not be constant at all rates of working.

In Table 6.1 the observations of Wiley & Newburgh (1931) are compared with calculations based on a mean skin temperature of 33.5 °C and a respiratory exchange of 6.0 l/min. The agreement

6. Respiration and insensible water loss

TABLE 6.1. *Calculated and observed rates of insensible water loss at rest at various humidities. Air temperature 28.0 °C, skin temperature 33.5 °C (assumed), respiratory exchange 6.0 l/min (assumed). r.h. ambient relative humidity; p_a, ambient water vapour pressure; $\dot{m}_w$, rate of water loss from the respiratory tract, from equation (5), expressed per unit surface area; $\dot{m}_{is}$, rate of water loss from the skin, from equation (7). The observed rate of water loss is taken from the experimental results of Wiley & Newburgh (1931).*

r.h. %	p_a kPa	$\dot{m}_w$ g/m².h	$\dot{m}_{is}$ g/m².h	Rate of water loss, g/m².h	
				Calculated	Observed
20	0.76	5.9	13.7	19.6	21.6
40	1.51	5.0	12.4	17.4	17.0
60	2.27	4.1	11.1	15.2	15.0
80	3.03	3.2	9.7	12.9	14.4

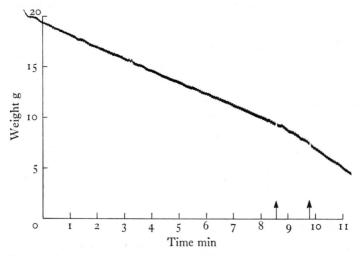

Fig. 6.5. Continuous record of weight (arbitrary zero) of a nude subject at rest in air at 21 °C. Between the arrows the subject did some mental arithmetic. This was signalled by interrupting the recording light beam. The rate of weight loss increased from 1.07 g/min (70 g/h) to 1.85 g/min (111 g/h). (From Kerslake, 1967b.)

is surprisingly close, since variations between subjects, and for one subject on different occasions, may be large (perhaps ± 50 per cent) (Winslow, Herrington & Gagge, 1937a, b; Newburgh & Johnston, 1942). These variations are probably mainly due to differences in pulmonary ventilation and in non-thermal sweating.

Obvious increases in palmar sweating occur with anxiety and other emotional changes, and Kuno (1934, pp. 131–4) found considerable increases when subjects performed mental tasks. The effect of mental arithmetic on the total rate of weight loss of a subject in a neutral environment is shown in Fig. 6.5.

Studies of sweating from the finger pads have demonstrated the effects of hormonal activity (Harrison & MacKinnon, 1962) and of the emotional reaction to surgical hospitalization (Harrison, MacKinnon & Monk-Jones, 1962). The diffusional loss of water through the epidermis also shows variations (Ohara, Kondo & Ogino, 1963). At the same skin temperature the rate of water loss from the non-sweating skin of the forearm is greater in summer than in winter (Heerd & Opperman, 1966). This may be due to seasonal changes in the epidermis and the sebaceous glands.

7 PHYSIOLOGICAL RESPONSES

Regulation of body temperature

It is customary to regard the body as composed of at least two compartments, a core, in which temperature appears to be controlled by the thermoregulatory mechanisms, and a shell of more superficial tissues where temperature is subject to greater variations. Changes in blood flow alter the conductance between core and shell, and sweating promotes loss of heat from the skin surface. The core must be regarded as a concept rather than as an anatomical entity, and core temperatures, measured at such accessible sites as the oesophagus, tympanic membrane, ear canal, mouth and rectum, usually differ from one another. However they behave in much the same way when work rate or environment is changed, although the time relations and magnitudes of the changes may vary from site to site (Piironen, 1970).

The classic work of Nielsen (1938) showed that over a wide range of environments the core temperature (in this case the rectal temperature) in the steady state is independent of the environmental stress, being closely regulated at a level which depends on the rate of working. This has since been confirmed many times (Winslow & Gagge, 1941; Robinson, 1949; Wyndham *et al.* 1952*a*, *b*; 1954; Astrand, 1952; Nielsen & Nielsen, 1962; Lind, 1962; 1963*a*, *b*; Blockley, 1965; Saltin & Hermansen, 1966; Webb & Annis, 1967; 1968; Stolwijk, Saltin & Gagge, 1968). Regulation fails if the environmental conditions are very severe (Fig. 7.1), and it is convenient to refer to conditions cooler than this as the prescriptive zone (Lind, 1963*a*). The zone is called prescriptive because it has been used as a criterion for setting limits for industrial work.

Nielsen (1938) proposed that a very efficient and sensitive mechanism controls the core temperature, and that the setting of this thermostat depends on the rate of working. This proposal was based on steady state observations in which the rate of heat production, closely related to the rate of working, was equal to the rate of heat loss. An alternative interpretation of the findings is that the rate of heat loss is controlled at a level which depends on the core temperature. These two descriptions are equally good so long as the rate of heat loss is identified by the rate of working. They

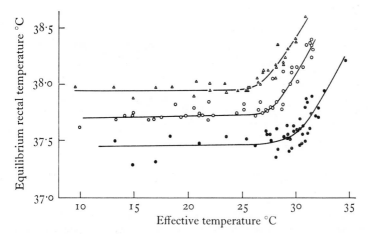

Fig. 7.1. Equilibrium levels of rectal temperature at three work rates in different environments. Rectal temperature is almost independent of environmental conditions over a wide range known as the prescriptive zone, but regulation fails when conditions are very severe. (From Lind, 1964.)

can be distinguished if the usual strong correlation between these quantities in the steady state can be broken.

Nielsen & Nielsen (1965 *b*) compared levels of rectal temperature in subjects either producing heat by working, or receiving the same amount of extra heat by diathermy at rest. Diathermy releases heat throughout the body tissues and this heat must be transferred to the surface in much the same way as heat of metabolic origin.

The rectal temperature in the steady state was elevated to the same extent by diathermy as by exercise, i.e. the elevation of core temperature depended on the rate at which body heat had to be eliminated rather than on the rate of working (Fig. 7.2). Consistent with this is the observation of Nielsen (1968) that the elevation of core temperature in exercise is the same whether the work is done by the arms or the legs.

7. Physiological responses

Another way of breaking the usual correlation between work rate and heat production is to compare the effects of positive and negative work (Nielsen, 1966; 1969). In positive work some of the metabolic energy appears as external work, so that the heat production within the body is less than the metabolic energy production.

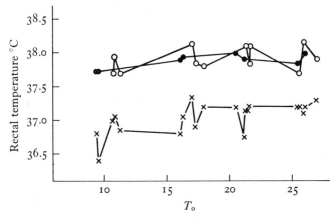

Fig. 7.2. Equilibrium levels of rectal temperature at different environmental temperatures, T_0. ×, at rest before the experiment; ●, producing additional heat by muscular work; ○, at rest with diathermy liberating heat in the tissues at the same rate as the muscular work. The elevation of rectal temperature depends on the total heat which the body must lose rather than the rate of metabolic heat production. (From Nielsen & Nielsen, 1965 b.)

In negative work the active muscle, instead of shortening, is stretched, and the energy required to do this is derived from the motor which drives the ergometer or treadmill. There is some doubt about whether the chemical engine can be driven backwards, but even if it can, most of the energy must appear as heat in the tissues, and the total heat liberated in the body during negative work must be greater than the metabolic energy production (Hill & Howarth, 1959; Wilkie, 1964).

The increase in core temperature was found to be more closely related to metabolic energy production than to the rate at which body heat had to be eliminated (Fig. 7.3). It is difficult to reconcile this with the results of the diathermy work. The only important difference appears to be the way in which extra heat was fed into the tissues, yet the results are essentially opposite. Nielsen (1969) draws attention to the possible role of muscle temperature as an

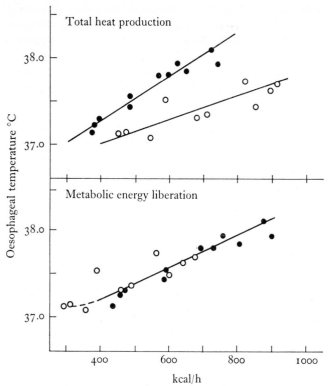

Fig. 7.3. Equilibrium levels of oesophageal temperature during positive (●) and negative (○) work (environmental temperature 22.3 °C). In the upper graph the temperature is plotted against the total heat liberated in the body (i.e. metabolic heat minus external work). Where the external work is negative the total heat is greater than the metabolic heat production. In the lower graph the temperature is plotted against metabolic heat production calculated from the oxygen consumption without adjustment for external work. (From Nielsen, 1969.)

input to the thermoregulatory control mechanism. If muscle blood flow is determined in part by metabolites produced by muscle activity, the tissue would run cooler in positive work than in negative work. The measured core temperature might therefore be raised to a higher level in positive work, when the muscles are cooler. The inference that muscle temperature is one of the inputs to the thermoregulatory system has also been drawn, on other evidence, by Robinson *et al.* (1965), Stolwijk & Hardy (1966*b*), Saltin & Hermansen (1966) and Saltin, Gagge & Stolwijk (1968).

5-2

7. *Physiological responses*

Saltin & Hermansen (1966) have compared the core temperature elevations (in the prescriptive zone) of subjects having different capacities for physical work. In Fig. 7.4(*a*) oesophageal temperature is plotted against oxygen uptake, and there are marked differences between subjects. These differences are greatly reduced when oxygen uptake is expressed for each subject as the percentage

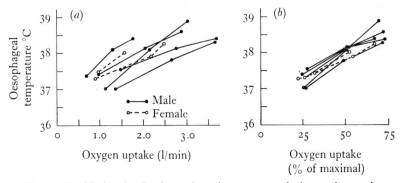

Fig. 7.4. Equilibrium levels of oesophageal temperature during work at various rates (environmental temperature 19–22 °C). The maximum oxygen uptakes of the two female (dashed lines) and five male (continuous lines) subjects ranged from 2.61 to 5.35 l/min. The differences between subjects are reduced when the oxygen uptake is expressed as a proportion of the maximum for each subject. (From Saltin & Hermansen, 1966.)

of his maximal oxygen uptake (Fig. 7.4(*b*)). The observations confirm and extend Astrand's (1960) finding that the same rectal temperature (38.1 °C) was reached by four subjects working at 50 per cent of their individual maximum oxygen uptakes (2.2 to 5.4 l/min).

The results shown in Fig. 7.4(*b*) can be summarized as follows

$$T_c = 36.5 + 3.0 \dot{V}_{O_2}/\dot{V}_{O_2 max}, \qquad (1)$$

where $\dot{V}_{O_2}$ and $\dot{V}_{O_2 max}$ are the actual and maximal rates of oxygen uptake. These subjects were men and women in the age range 24–30 years, when maximum oxygen uptake is highest (Webb, 1964). The same relation probably holds for older and younger subjects (Astrand, 1960). For young men (maximum oxygen consumption 4 l/min) the core temperature in the steady state in the prescriptive zone can be expressed in terms of heat production per unit area, H_p,

$$T_c = 36.5 + H_p/250. \qquad (2)$$

A relation of this form would result if some layer in the body between the core and the skin were maintained at 36.5 °C under all conditions, the conductance between this level and the core being constant. It is as though inside the homoiotherm there is an archetypal poikilotherm who lives in a constant environment. Unhappily for this attractive picture there does not appear to be a level in the tissues where the temperature is regulated in this way under all conditions of work and environment in the prescriptive zone.

Non-steady states

If a subject in the prescriptive zone changes his rate of working, his core temperature will eventually reach the level indicated by equation (1). One would expect a sensitive thermostatic mechanism which was reset according to his rate of working to initiate responses tending to raise the core temperature rapidly to the new level. Instead of this, sweating increases almost as soon as work begins (van Beaumont & Bullard, 1963), retarding the rise in core temperature.

The responses to a change in work rate are shown in Fig. 7.5. The core temperature reaches its new equilibrium level after about 20 min (Nielsen & Nielsen, 1965 a). A subject having a high working capacity works harder in order to reach a certain temperature plateau (equation (1)) but his initial rate of change of core temperature is similar to that of subjects with smaller working capacities working at correspondingly lower rates (Saltin & Hermansen, 1966). If his thermostat is indeed reset to a new level, one might expect him to make use of his greater heat production to attain the target level more rapidly. In order to explain these findings in terms of a thermostatic mechanism it is necessary to suppose that the deployment of vasodilation and sweating varies with the discrepancy between actual core temperature and its set point over a rather wide band. This then introduces complications when one attempts to account for the close control of core temperature at constant work rate in different environments (Fig. 7.1). Further inputs from skin and muscle must be introduced in order to explain this (Hammel, 1968) and the concept of the thermostat becomes so overlaid that its utility is doubtful (Brown & Brengelmann, 1970).

The results shown in Fig. 7.5 are consistent with the proposition

7. Physiological responses

that the rate of heat loss is controlled at a level depending on the core temperature. The equation for the rate of heat storage, H_σ, at any time, may be derived from equation (1). For fit young men (equation (2)) it is

$$H_\sigma = H_p - 250(T_c - 36.5).\qquad\qquad(3)$$

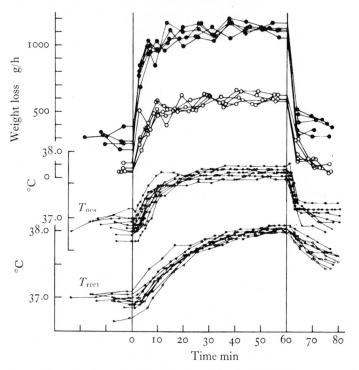

Fig. 7.5. Rate of weight loss, oesophageal and rectal temperature before, during and after 60 min work at 900 kpm/min (147 W). ○, $T_0 = 20$ °C; ●, $T_0 = 37.5$ °C. (From Nielsen & Nielsen, 1965 a.)

If the stored heat were shared equally throughout the body it would be possible to find the rate of change of core temperature by dividing H_σ by the heat capacity of the body per unit area. Core temperature would then approach equilibrium exponentially with a time constant of about 9 min. The assumption that stored heat is distributed uniformly is clearly false. The temperature of working muscle increases more than core temperature and skin temperature is almost independent of work rate (Nielsen, 1969).

At the beginning of exercise, heat storage must be predominantly in the muscles, and the increase in core temperature should be slow. When exercise stops one would expect the fall in core temperature to begin almost at once since the resting muscles do not actually remove heat. These features are evident in Fig. 7.5, and the temperature reaches a plateau rather sooner than would be expected if heat were stored uniformly. In the same experiments the rectal temperature took about 60 min to reach a plateau, showing that the steady state was not reached for at least an hour despite constancy of the oesophageal temperature during the last 40 min. In similar experiments on another subject (Nielsen & Nielsen, 1965 a) oesophageal temperature again reached a steady level after about 20 min, but sweat rate increased throughout the whole hour.

Identification of oesophageal temperature with the controlled or controlling core temperature may be too naive. If muscle and other temperatures are involved the increase in heat loss after oesophageal temperature has reached a plateau can be explained, as can the discrepancy in the case of negative work.

More severe environments

The relation between core temperature and heat loss which holds over the prescriptive zone alters when conditions are very hot (Fig. 7.1). As ambient temperature and humidity increase, a point will be reached when the skin temperature consistent with the required rate of heat loss is such that an increase in core temperature will have to occur if the internal transport of heat to the skin is to be maintained. However Lind (1964, pp. 99–101) has pointed out that at low rates of working the core temperature begins to rise considerably before this physically limiting point is reached, and at submaximal levels of sweating and skin blood flow. Perhaps at a given core temperature only a limited deployment of peripheral blood flow and sweating can occur, this maximum deployment increasing at higher core temperatures.

The central control of sweating

The core temperature of a subject producing 200 W/m² heat is almost independent of ambient temperature over the range 20–40 °C if humidity is low (Fig. 7.1). Over this range of ambient

7. Physiological responses

temperature the evaporative requirement for heat balance changes by about 150 W/m² (assuming that $h_0 = 10$ W/m².°C and the skin temperature varies from 31 to 36 °C). If the same subject in a fixed environment decreases his heat production, and therefore his evaporative requirement, by 150 W/m², his core temperature (oesophageal, rectal or tympanic) will fall by about 0.6 °C (equation (2)). The control of sweat rate cannot depend solely on core temperature, although such a system has been proposed (Benzinger, 1959; 1961).

A strong correlation between sweat rate and core temperature is to be expected if the work rate is altered in a constant environment. There is little change in skin temperature in such a case, so the evaporative requirement increases directly as the heat production, as does the core temperature (equation (1)). In environments hotter than the prescriptive zone, at the right-hand end of Fig. 7.1, the core temperature at constant work rate increases with the environmental stress, and here a correlation between core temperature and sweat rate is to be expected even at constant work rate. No such relation with core temperature is seen in the prescriptive zone at constant work rate, when core temperature is almost independent of environmental conditions.

Peripheral inputs

In order to account for changes in sweating at constant core temperature it is necessary to propose peripheral inputs to the heat regulating centres since the effect of local skin temperature on sweat gland activity is too small to explain the large changes which can occur (p. 142). The existence of peripheral inputs to the central sweat control mechanism is well established (Randall *et al.* 1963; Robinson, 1965). Rapid cooling of a small area of skin, as by application of an ice pack, causes general inhibition of sweating (Kuno, 1934; Randall, Deering & Dougherty, 1948; Issekutz, Hetenyi & Diosy, 1950; Rawson & Hardy, 1967; Banjeree, Elizondo & Bullard, 1969). The inhibition is frequently transient (Hill, 1921) as it must be if the rate of removal of heat from the cooled area is insufficient to maintain body heat balance. Sweating is again inhibited if the ice pack is moved to another skin region. The response is thought of as transient because it passes off even though the ice pack remains in contact with the skin. This does not necessarily imply adaptation in the neural mechanism, since skin temperature itself may not be the important afferent stimulus. If this stimulus were the rate of change of skin temperature (Wurster & McCook, 1969), a further time derivative, the rate of heat loss, or temperature gradients in the skin (Bazett, 1951), it would be very large when the ice pack was first applied, but would diminish thereafter as the tissue temperatures and temperature gradients settle down. A mechanism depending on time derivatives of skin temperature would be without effect in the steady state, whereas one depending on skin heat loss or temperature gradient would operate at all times. If the rate of skin heat loss were the stimulus, the effect would be to inhibit an important heat loss mechanism at high rates of heat loss, i.e. at high work rates in the steady state. This may appear undesirable, but such a mechanism would have inherent stability and could explain the elevation of core temperature associated with physical work.

Local or general application of a warm stimulus causes a general increase in sweat production (Kuno, 1934; Rawson & Randall, 1961; Gagge & Hardy, 1967). The response to heating the skin over the spine may be great enough to lower the core temperature (Roe, Hardy & Stolwijk, 1967).

The change in sweat rate is very rapid. Fig. 7.6 shows a continuous record of the weight of a nearly nude subject. The oscillations are caused by respiration and give some indication of the sensitivity and response speed of the balance. When radiant heaters were switched on for one minute there was a rapid increase in the rate of weight loss, which appeared to reach a new steady level within about half a minute. When the heaters were switched off the rate of weight loss returned nearly to its previous level within half a minute. Changes in sweat production must have been even more rapid than the weight record indicates, since a change in evaporation requires a change in wetted area, and some accumulation of unevaporated sweat is needed to bring this about. When the heaters were switched on there was a sudden change in the rate of heat loss from the skin. This was almost exactly countered by the sweating response so that the net heat loss of the subject was soon restored almost to its previous level.

Rapid changes in general sweat production can also be examined by measuring the sweat output from a small area of skin by means of a ventilated sweat capsule. Spontaneous fluctuations in sweating from different regions are synchronous (Albert & Palmes, 1951;

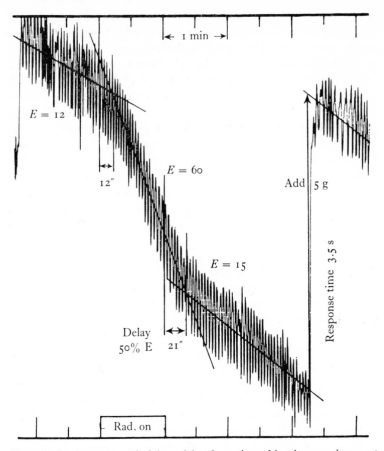

Fig. 7.6. Continuous record of the weight of a resting subject in an environment at 30 °C. Radiant heaters were switched on for 1 min where indicated (Rad. on), raising the Operative temperature to 43 °C. (Unpublished data observed in connection with the work of Gagge & Hardy (1967).)

Randall *et al.* 1965), and the time relations of the sweat responses in one region are probably representative of those in others (Custance, 1965; Brebner & Kerslake, 1969). Fig. 7.7 shows changes in the forearm sweat rate in response to radiant heating of the chest for 30 s. Readings are at intervals of 5 s; the time constant of the gas circuit and analyser was about 5 s, with an absolute delay of 6 s. The increase in sweating begins within 2 s and the curve suggests that the sweat rate might become constant after about 1 min heating. The small peak about 40 s after heating was stopped was observed in all the runs and may indicate excessive inhibition of sweating at the beginning of the cooling period.

When the intensity of radiant heat to the chest is varied sinusoidally (period 8–70 s), the forearm sweat response, averaged over a number of cycles, is approximately sinusoidal, the peak

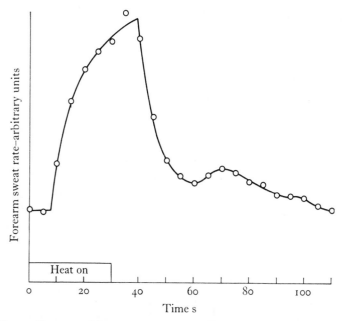

Fig. 7.7. Response of forearm sweat rate to radiant heating of the chest for 30 s. Mean of ten experiments on one subject. Forearm sweat rate was measured with a ventilated sweat capsule and infra-red gas analyser.

evaporation rate occurring about 1.6 s after the peak radiation intensity at all frequencies (Brebner & Kerslake, 1961 *b*). One would expect a phase lag between the radiation intensity and the skin temperature of the heated area, i.e. the interval between peak radiation and peak skin temperature should vary with the frequency of the stimulation cycle. There is no evidence of such a phase lag in the sweat response, and it seems probable that the receptors in the heated area are sensitive either to the rate of change of skin temperature or to the rate of heat loss. The effect of skin temperature itself must be small.

7. Physiological responses

The results shown in Fig. 7.8 are of great interest in this context. A nearly nude subject was continuously weighed while exposed to radiant heat at various intensities. The experiment is divided into two sections. In the first, heat was applied for 3 min (B), switched off for 3 min (C), applied again at a different intensity

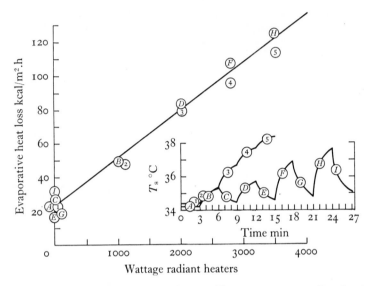

Fig. 7.8. Response of sweat rate and mean skin temperature to radiant heating. The inset curve shows the sequence of the experiments. Increasing intensities of radiant heat were applied for 3 min each, either with or without 3 min periods during which the radiation was switched off. The skin temperatures in the two sequences are quite different, but the sweat responses are similar. (Unpublished data observed in connection with the work of Gagge & Hardy (1967).)

for 3 min (D), switched off again (E) and so on. In the second, the heat was not switched off but was increased to a new intensity every 3 min. The sweat rate at a given intensity of heating was the same regardless of the previous condition, and was nearly sufficient to dissipate the extra heat gained by radiation. The skin temperatures, on the other hand, were quite different. In the nature of things there must be a relation between the level of skin temperature and its rate of change in each radiant condition, and it could be proposed that the sensitivity of the skin receptors is such that exactly appropriate weight is given to skin temperature and to its rate of change (Wurster & McCook, 1969). A conditioned reflex mechanism might help to explain the accurate anticipation of the requirements for heat balance. However, this is hard to reconcile with the results of sinusoidal stimulation, where skin temperature appeared to have no significant effect. Both sets of results are more simply explained on the assumption that the receptors are sensitive to the rate of heat loss and that this is accurately controlled by the heat regulating mechanism.

Experiments on the cooling of occluded limbs (Brebner & Kerslake, 1961 *a*) suggest that the response of the receptors may depend on temperature differences between cutaneous vascular plexuses, as Bazett (1951) has proposed. Evaporative cooling of the unoccluded limb produced general inhibition of sweating, but the response was abolished if the circulation to the cooled limb was occluded, despite the fact that the limb then cooled faster and felt colder. In contrast, radiant heating of occluded limbs produces a general sweat response. This may be because radiation which penetrates the epidermis is absorbed more in the venular than in the arteriolar plexuses thus increasing the response to radiant as opposed to convective or evaporative stimuli (Kerslake, 1968).

If the subject is not sweating before a warm stimulus is applied, the sweat response to heating is much slower (Adolph, 1946; McCook, Wurster & Randall, 1965). On exposure to moderate heat (environment at 30–40 °C) there is a small increase in evaporative heat loss in the first 5 min, but this may represent only a change in insensible perspiration and loss of water from the upper epidermis resulting from the rise in skin temperature. Custance, Heath & Cattroll (1970) consider that it is due to sweat gland activity. There is usually a delay of 10–30 min before sweating begins, and the sweat rate than increases only slowly (Colin & Houdas, 1965). Timbal *et al.* (1969) found that after sweating had begun it increased exponentially to its final equilibrium value. A time constant of 11 min is quoted for one experiment (other values are not stated). When subjects are exposed to hotter environments (above 45 °C) there is frequently no observable delay in the onset of sweating (Belding & Hertig, 1962; Hardy & Stolwijk, 1966).

7. *Physiological responses*

The effect of local conditions on sweat rate
Skin temperature

Local skin temperature can affect local sweat production in two ways; by directly stimulating glands which would otherwise be inactive, and by modifying the rate of secretion of glands which are receiving sudomotor stimuli of central origin.

Direct stimulation of glands which are not receiving central sudomotor stimuli was described by Saito (1930) and has been confirmed under different conditions by Gurney & Bunnell (1942), Randall (1947), Issekutz, Hetenyi & Diosy (1950) and Bullard, Banjeree & MacIntyre (1967). Such stimulation is rarely seen at skin temperatures below 40 °C, and is probably best regarded as a local protective mechanism. Kuno (1956, pp. 284–91) has suggested that it depends on the axon reflex system which is known to be associated with the innervation of the sweat glands (Coon & Rothman, 1941; Collins & Weiner, 1961), but McLaughlin & Sonnenschein (1963) point out that the response is very closely limited to the area heated, differing in this respect from other responses known to involve the axon reflex mechanism. The response is present in denervated glands (Rothman & Coon, 1940) and is not abolished by local injection of procain (Benjamin 1953).

Local heating to temperatures below 40 °C can cause local sweating where none was present before (Benjamin, 1953), but this occurs only in warm ambient conditions when central sudomotor impulses may be reaching the glands. Lloyd (1959) has shown that at low rates of neural stimulation glands may be active but the secreted sweat may not reach the surface. Raising the local temperature might increase the output of the glands in response to such stimulation sufficiently to cause sweat to appear on the surface.

The effect of local skin temperature on the output of sweat glands which are receiving sudomotor stimuli of central origin has received surprisingly little attention in the past. The existence of any significant effect at skin temperatures higher than 33 °C was denied by Benzinger (1959; 1961), who claimed that the rate of sweat production under these conditions depends solely on the hypothalamic temperature.

In considering the effect of skin temperature on sweat rate it is necessary to distinguish between an effect of general skin temperature on the central drive to the sweat glands and an effect of local temperature on the response of the glands to this drive. In order to exclude the former, symmetrical skin regions (preferably small) can be maintained at different temperatures. The central drive may be influenced by this, but if the reasonable assumption is made that the distribution of central drive is not modified by local temperature, differences in the sweat rate in the two regions can be ascribed to the effect of local temperature on the response of the glands at the two sites to equal neural stimuli.

Robinson *et al.* (1950*a*) imposed different skin temperatures on the two forearms and hands, which were enclosed in bags. Surprisingly, the cooler arm (30.2 °C) sweated more on average than the warmer (36.9 °C). More severe cooling to 13 °C was found by Fox *et al.* (1964) to reduce sweating to about 10 per cent of that on the control side, where the skin temperature was 37.5 °C. Neither of these investigations was primarily concerned with the effect of local temperature on sweat rate, and the conditions of wet skin within the arm bag considerably complicate interpretation of the findings. Not only is sweating reduced when the skin is wet (hidromeiosis), but the skin of the hand is able to take up a good deal of water. This may eventually pass back into the blood stream as it does in the case of the foot (Folk & Peary, 1951; Buettner, 1959*a*).

The problem has been approached directly by Bullard, Banjeree & MacIntyre (1967), using ventilated capsules on the thighs. The skin under one of the capsules was heated or cooled by heating or cooling the capsule, and sweat output compared with that from the control capsule where the skin temperature was held constant. With this technique it was possible to measure the effect of local temperature at different levels of central drive produced by exposing the subject to different environments. At all levels of central drive local sweating increased with local skin temperature. For given skin temperatures under the two capsules the ratio of the sweat rates on the two sides was much the same whatever the level of central drive. Local skin temperature thus appeared to act as a multiplier of central drive. At skin temperatures below 35 °C the increase in sweating with skin temperature was consistent with a Q_{10} of between 2 and 3; above 35 °C with a Q_{10} of between 4 and 5. When local sweating was induced by pilocarpine or inhibited by

7. Physiological responses

atropine (applied by electrophoresis) the effect of local skin temperature was less. It appears that the main effect of temperature is on some part of the system linking neural impulses of central origin with stimulation of the glands. This conclusion is supported by the work of Ogawa (1970).

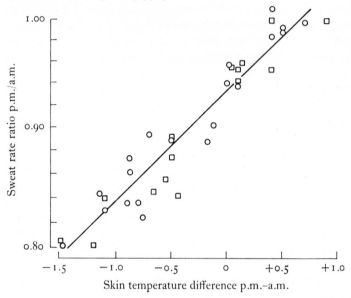

Fig. 7.9. Effect of skin temperature on maximum sweat production of two subjects at rest. Maximum sweat rate was measured each morning and afternoon. Differences in mean skin temperature were produced by altering the temperature of the saturated environment in which the subject was weighed after previous heating to a core temperature above 38 °C. The ordinate shows the ratio of the sweat rate in the afternoon experiment to that in the morning, the abscissa the difference in the temperature. (From Kerslake & Brebner, 1970.)

Another way of removing the effect of variations in central drive is to work under conditions in which this is maximal. Kerslake & Brebner (1970) measured the total sweat rates of subjects in whom the central drive appeared to be maximal. After adjustment for hidromeiosis (see below) the sweat rates were steady despite rising core temperature. Experiments were performed on the morning and afternoon of each day, mean skin temperature being different on the two occasions. In Fig. 7.9 the ratios of the sweat rates in the afternoon and morning are plotted against the differences in skin temperature. The slope of the line indicates the effect of skin temperature and over this range (36.7–39.2 °C) is consistent with a Q_{10} of 2.9. It will be noticed that when the skin temperature was the same on both occasions the sweat rate in the afternoon was only about 93 per cent of that in the morning. Conditions in the morning experiment were nearly constant from day to day, and whether this reduction in sweating in the afternoon is a diurnal effect or a consequence of the stress of the morning experiment it is likely to be a constant factor which does not affect the interpretation of the effect of skin temperature.

Hidromeiosis

Etymologically the term hidromeiosis merely means a reduction in sweating, but it has become restricted to mean a particular type of reduction in sweating associated with wetting the skin (Sargent, 1962). Hydrohidromeiosis might be a more appropriate word. In 1956, Peiss, Randall & Hertzman showed that sweating from the palm and fingers could be suppressed by immersing the hand in water or dilute saline. Strong saline did not produce the effect, which was most marked when distilled water was used. They suggested that swelling of the cells lining the sweat duct might produce mechanical obstruction and so prevent the sweat escaping. It was later shown that the suppression was greater in warm water (25–40 °C) than in cold water (5–15 °C). In 10 per cent and 20 per cent sodium chloride solution it was slight or absent (Randall & Peiss, 1957).

The whitening and wrinkling of the skin which occurs when the hand is soaked in water draws attention to the epidermal swelling and makes the explanation of poral closure an attractive one. In regions other than the palm and sole the epidermis is thinner and the swelling less obvious. Perhaps for this reason, and because blockage of sweat ducts is frequently accompanied by urticaria (Sulzberger *et al.* 1953), the possibility of a similar symptomless effect on the general body surface was not directly investigated for several years. It was well known that during exposure to heat the sweat rate often declines after an hour or two (Ladell, 1945; Gerking & Robinson, 1946; Thaysen & Schwartz, 1955), but this was usually interpreted as fatigue of the sweat glands or of the controlling mechanisms. Fatigue should be a direct consequence of activity, and one would expect it to be more evident at high sweat rates. This was broadly true, but Gerking & Robinson (1946)

7. *Physiological responses*

found that the decline in sweating was more marked in humid environments, when unevaporated sweat was dripping off the body. Here one might postulate an adaptive mechanism whereby the central drive might be reduced in order to conserve water. Taylor & Buettner (1953) also suggested that sweat production was related to the evaporative capacity of the environment.

When an arm is enclosed in an arm bag during a heat exposure the sweat rate of the arm at first rises with that of the general body surface but later diminishes progressively even if the general sweat rate is maintained (Fig. 7.10). Collins & Weiner (1962 a) investigated this suppression of arm bag sweating and concluded that it must be related to wetting of the skin surface with sweat. They examined the possibility that water might be reabsorbed through the skin (van Heyningen & Weiner, 1952 a) but found that the rate at which this might occur was too small to account for the observed decline in sweating. Sweating from the arm could be inhibited by soaking in water for several hours and the inhibition was greater the higher the temperature of the water. It did not occur when the arm was soaked in 10 per cent sodium chloride solution. In these respects the phenomenon was similar to that described by Randall & Peiss (1957) in the palm, and poral closure as a result of epidermal hydration seemed the most likely explanation (Collins & Weiner, 1962 b). This receives further support from the fact that hidromeiosis is a strictly local effect which can be prevented or reversed by stripping off the outer layers of epidermis (Brown & Sargent, 1965).

Collins & Weiner (1962 a) suppressed local sweating by repeated injections of atropine into a small region of the skin of an arm enclosed in an arm bag. There was some sweating from the remainder of the limb, and the skin was wet. After four hours the sweat response of other regions of the arm to local injection of acetylcholine had diminished, but where sweating had been inhibited the response was normal. This confirms the earlier observations of Thaysen & Schwartz (1955) and suggests that hidromeiosis occurs only in regions which are actively sweating, a conclusion at variance with the findings of Hertig, Riedesel & Belding (1961) for the whole body, but supported by those of Brebner & Kerslake (1968).

The decline in sweating from the general body surface was investigated by Hertig (1960) and Hertig, Riedesel & Belding (1961).

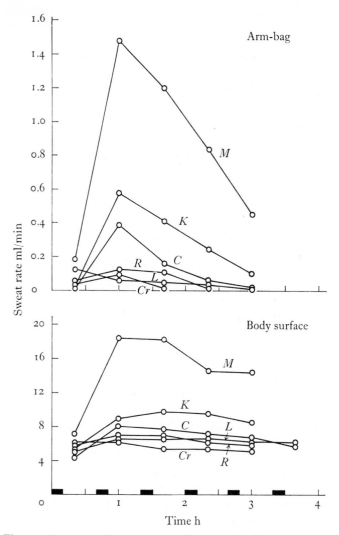

Fig. 7.10. Sweat rates from the arm and hand, enclosed in an arm bag, and from the general body surface. Six subjects, identified by letters in the figure, performed a work and rest routine in a hot environment ($T_o = 40$ °C, $p_a = 4.2$ kPa). The blocks on the abscissa indicate work periods. After the first hour there is a marked decline in sweat production from the arm, while general sweat production falls only slightly. (From Collins & Weiner, 1962*a*.)

7. Physiological responses

Subjects were immersed for several hours in stirred water at controlled temperature. Sweat loss was measured by removing and weighing them at intervals. Steady levels of core temperature were reached by the end of the first hour, but sweat rates subsequently declined exponentially towards zero. This decline did not occur if immersion was in 10 per cent saline. The exponential nature of the decline has been confirmed by Fox et al. (1963) and Brebner & Kerslake (1964). Both these investigations showed that hidromeiosis was established within 15 min of immersion.

The decline of sweating is described by the equation

$$S_t = S_0 . e^{-t/k}. \tag{4}$$

Here S_0 is the sweat rate that would obtain in the absence of hidromeiosis under the prevailing conditions of local skin temperature and central sweating drive. Experimental determinations of the time constant, k, are based on measurements made under conditions in which S_0 is constant. In Hertig's work the skin temperature and core temperature were constant and it is plausible to assume that S_0 was constant. Brebner & Kerslake (1964; 1968) worked at core temperatures above 38 °C with almost no exchange of heat at the skin surface, when there is reason to believe that the sweating drive is maximal (Kerslake & Brebner, 1970). Time constants for their three subjects lay between 90 and 140 min. Fox et al. (1969) examined a group of British men before and after a course of heat acclimatization, and groups of Swedish men and women. The subjects were maintained at constant core temperature and were therefore roughly in heat balance, with skin temperature lower than core temperature. Sweat production was measured by sucking the sweat out of an impermeable bag enclosing the subject (Fox, Crockford & Löfstedt, 1968). There is some evidence that sweating drive may not be quite maximal in this condition, and that variations in heat loss which were imposed in order to regulate core temperature might affect the central drive (Kerslake & Brebner, 1970). The results are interpreted on the assumption that sweating drive was constant. Sweating declined significantly faster in the unacclimatized men ($k = 60$ min), but there was no significant difference between the other groups of men ($k = 90$ min for acclimatized British men and 100 min for Swedish men). The women showed only a slight decline in sweating, but there is some evidence that their sweat rates continued to rise during the early part of the experiment, implying that central drive was not constant. Their sweat rates were so low (about 30 g/15 min) that the method of collection may have introduced important errors.

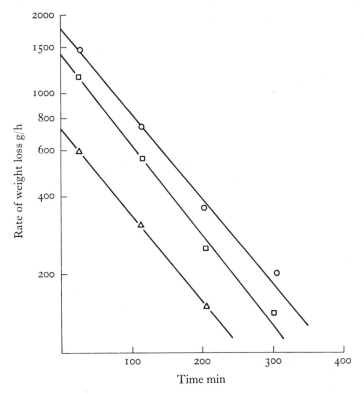

Fig. 7.11. Decline in maximum sweat production of three subjects during prolonged immersion in water. For most of the time the subject was immersed in water at 36 °C. Before each measurement of maximum sweat production he was heated to a mouth temperature of 38.5 °C and then transferred to an environment of saturated air at 37 °C, where the measurement was made. (From Brebner & Kerslake, 1964.)

The conclusion that hidromeiosis proceeds more rapidly in men than in women is not supported by the observations of Wyndham, Morrison & Williams (1965). Male and female subjects worked for 4 h in a humid environment, $T_0 = 33.9$ °C, $p_a = 4.71$ kPa, $V = 0.41$ m/s. From the third to fourth hours the mean sweat rate of the male group fell from 524 to 480 g/m².h, and that of the female group from 344 to 195 g/m^2. h. The reduction of sweating in the females was thus both absolutely and relatively greater than that in the males. Both groups were near thermal equilibrium. The mean rectal temperature of the male group increased about 0.1 degC during the last hour and that of the female group fell about 0.1 degC. The decline in sweating in the male group is smaller than one would expect on the assumption that central drive was constant and the skin completely wet. On the basis of an exponential decline towards zero the time constant would be about 10 h. The results for the female group are consistent with a time constant of 105 min. It is probably unjustifiable to interpret the observations so simply, both because the central drive was probably changing in both groups and because some evaporation of sweat was taking place.

In contrast to the slow development of hidromeiosis when the skin is wet, recovery is very rapid when the skin is allowed to dry. Brebner & Kerslake (1968) found that recovery of sweating over the whole body was about 50 per cent complete after 5 min exposure to an environment at 33 °C, $p_a = 2.0$ kPa, $V = 0.25$ m/s. A comparable rate of recovery was observed for the forearm by Collins & Weiner (1962a). For palmar skin the recovery takes considerably longer (Randall & Peiss, 1957), possibly because there is a much greater uptake of water by the epidermis per unit area of skin.

In summary, hidromeiosis presents the following features:

1. It is a local phenomenon.
2. The epidermis must be hydrated and swollen.
3. Sweating must be present.
4. The decline in sweating is exponential, tending towards zero.
5. Recovery is rapid.

The simplest explanation is that epidermal cells swell and block the orifices of the sweat ducts. This is inadequate on two grounds. It fails to explain why the ability to sweat does not decline when the skin is kept wet but the sweat glands are inactive (indeed it would predict a decline), and it does not account for the exponential nature of the decline. Even if the epidermal swelling proceeded exponentially it seems likely either that all the sweat ducts would become obstructed after a finite time or that some would never become obstructed. In all subjects examined the decline appears to be towards zero sweating after infinite time, and the exponential law is closely followed.

An alternative explanation is that blocking of the sweat duct occurs when a fragment of solid material becomes detached from the secretory part of the sweat gland and impacted in the keratin ring at the mouth of the sweat duct. The cytoplasm of the secretory cells of the eccrine sweat glands is known to form blebs which project into the lumen. This arrangement is common in apocrine glands where these protrusions are thought to break off and contribute to the secretory process (Montagna, 1962, p. 410). The keratin ring (O'Brien, 1947) normally allows such fragments to pass through it, but becomes swollen when the epidermis is hydrated. It may be supposed that this swelling occurs relatively quickly when the skin is immersed in water. Liquid sweat can continue to pass through the constricted ring until it is blocked by a cellular fragment. Detachment of a cellular fragment in an eccrine gland may be supposed to occur somewhat rarely, at random times and only when the gland is active. If the keratin rings are swollen, blockage of the glands will proceed in a manner akin to atomic disintegration. At any time the probability of a patent duct becoming obstructed is constant, so the rate of blockage is proportional to the number of glands which remain patent. The decline in sweating will be strictly exponential, the sweat rate approaching zero at infinite time. If sweating is absent the glands will not become blocked. If the skin is allowed to become drier the keratin rings will shrink, releasing the impacted cell debris. Recovery from hidromeiosis will therefore be rapid. Stripping the outer epidermal layers will remove the keratin rings and so prevent or reverse blockage of the ducts. Immersion in dilute saline, as opposed to water, would produce less swelling of the keratin rings. Only large cellular fragments would then produce blockage, so the time constant of the decay in sweating would be larger, as Hertig *et al.* (1961) have found.

The degree of epidermal hydration required to produce hidromeiosis may not be the same as that required to maintain it, and neither of these quantities has been established. Immersion in 15 per cent sodium chloride solution does not produce hidromeiosis (Hertig *et al.* 1961). The relative humidity of this solution is about 0.92, so the skin relative humidity required to produce hidromeiosis probably lies between 0.92 and 1.0. This rather small

7. *Physiological responses*

range of relative humidity corresponds with quite a large change in epidermal water content. Fig. 7.12 shows the relation between the relative humidity of cornified epithelium and its water content. Between relative humidities of 0.5 and 0.9 the results have been fitted by a theoretical equation for multilayer adsorption developed by Brunauer, Emmett & Teller (1938). Agreement fails at higher values, the region of interest here. Extrapolation of the curve is

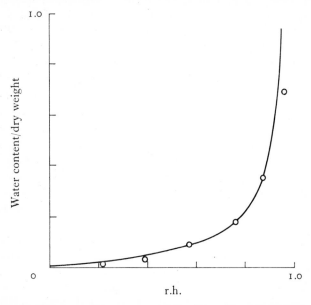

Fig. 7.12. Relation between the relative humidity of cornified epithelium and its water content (observations of Blank, 1952). The curve is a multi-layer adsorption isotherm (Brunauer, Emmett & Teller, 1938) fitted to three of the points.

hazardous, but it seems likely that the water content doubles between relative humidities of 0.9 and 1.0, and the critical point for partial or complete poral closure might well lie within this range. There is some evidence that the degree of epidermal hydration required to maintain blockage of sweat ducts is less than that required to produce it. Brebner & Kerslake (1964) found that after some degree of hidromeiosis had been produced by water immersion, a period of immersion in 15 per cent saline neither increased nor decreased the sweating capability. Hidromeiosis appeared to be frozen, suggesting that sweat ducts which were already blocked remained so, while no new blockages occurred. Peter & Wyndham (1966) observed progressive suppression of sweating in areas of skin which did not glisten with sweat. Their subjects were in a warm humid environment, $T_0 = 34$ °C, $p_a = 4.7$ kPa. In the absence of sweating the skin relative humidity would have been at least 0.8, but some glands were still active so the value must have been higher. As there was no loss of sweat by dripping, the mean salt content of the surface moisture must have been increasing towards saturation, but in the lumen and immediate vicinity of the patent sweat ducts it may have remained low. If the ducts which are already blocked do not unblock at a skin relative humidity of 0.82 (corresponding with saturated salt solution) hidromeiosis could proceed until all the ducts were blocked. In the environment used there would be no evaporation from saturated salt solution at 36 °C, so the skin surface would remain moist. At lower ambient water vapour pressures the saturated solution would dry out and the epidermis might shrink sufficiently for sweating to be re-established. As soon as some sweating began the skin would become moist again, preventing any further increase in sweating. The few patent ducts would be liable to blockage and the mean sweat rate would be such as to maintain a film of saturated salt solution on the skin surface.

The simple exponential decay towards zero which occurs when the skin is kept completely wet with water is not to be expected when subjects are exposed to an environment in which evaporation is possible. If the central drive is constant, sweating will decline in those areas where it initially exceeds the local maximum evaporative capacity. This decline will at first proceed exponentially towards zero, since the skin is covered with a film of sweat of very low salt concentration. Later, as sweating diminishes, salt will accumulate on the surface and the time constant of the decline in sweating will increase. Eventually the sweat rate from these initially wet regions will become constant, at a level equal to $h_e(0.82p_s - p_a)$. If p_a exceeds $0.82p_s$, the rate will become zero. In the regions where the sweat rate is initially less than the maximum evaporative capacity, sweating will not diminish.

The general decline in sweating will depend on the distribution of sweat rate and maximum evaporative capacity over the body surface, and its time course is unlikely to follow any simple law. The asymptotic level will be somewhat less than the mean maxi-

7. Physiological responses

mum evaporative capacity. If all regions are initially wet it will be given by

$$S = E_{\max} - 0.2h_e \cdot p_s. \tag{5}$$

In hot room experiments the picture is further complicated by changes in sweating drive, but experiments of several hours duration suggest that sweat rate probably does decline towards a level somewhere near the maximum evaporative capacity. Brown & Sargent (1965) observed the decline in total sweat production of subjects exposed to warm humid and hot dry environments. The humid conditions, $T_0 = 33\,°C$, $p_a = 4.0$ kPa, $M = 193$ W/m², were such that thermal balance was not reached. Body temperature rose throughout the eight hour exposure, while sweat rate fell from the second hour onwards. Subjects transferred after 6 h to a hotter, drier environment ($T_0 = 42\,°C$, $p_a = 2.5$ kPa) showed a rapid increase in sweating, indicating recovery from hidromeiosis. The development of hidromeiosis was related to the climate and the sweat rate. Time constants calculated on the assumption that sweating declined exponentially towards zero were larger in the drier environment and in subjects with low sweat rates. By extrapolation it was inferred that hidromeiosis would be absent in the humid environment if the initial sweat rate were below 432 g/h. The corresponding threshold for the drier environments was 1350 g/h. If an exponential decay is to be assumed (and there is no sound basis for this when evaporation is present) it must be supposed to tend to some finite sweat rate S_∞, after infinite time rather than to zero. Time constants should be based on plots of $\log(S - S_\infty)$, not $\log S$, against time. The latter plot will not be linear, but in so far as a time constant can be inferred it will increase as S_∞ increases. The hidromeiosis 'thresholds' are better regarded as estimates of S_∞. A very rough check of equation (5) can be made on the assumption that subjects were fully wetted at some stage. Neglecting the (likely) differences in skin temperature in the two environments

$$\frac{h_e(0.8p_s - 4.0)}{h_e(0.8p_s - 2.5)} = \frac{432}{1350},$$

$$p_s = 5.88 \text{ kPa}, \quad T_s = 35.8\,°C.$$

This is a realistic value for skin temperature under these conditions. Had it been assumed, the ratio between the values of S_∞ for the two environments as calculated from equation (5) would have corresponded exactly with the observations.

Robinson & Gerking (1947) measured the decline in sweating of subjects in a very humid environment,

$$T_0 = 34.5 \,^{\circ}\text{C}, \quad p_a = 5.05 \,\text{kPa}, \quad V = 0.92 \,\text{m/s}.$$

The average skin temperature of the three subjects was 35.8 °C (p_s = 5.88 kPa), so the relative humidity of non-sweating skin would have been 0.86. In such circumstances one would expect the

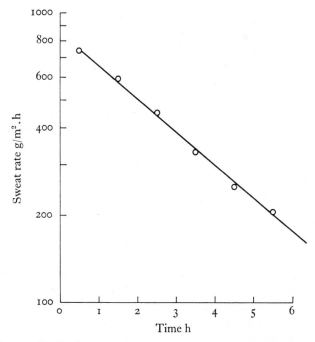

Fig. 7.13. Decline in sweating of subjects working in a very humid environment, $T_0 = 34.5 \,^{\circ}\text{C}$, $p_a = 5.05 \,\text{kPa}$, $V = 0.92 \,\text{m/s}$. Mean sweat rate for three subjects is plotted on a logarithmic scale. The time constant of the decline is about 3.8 h. (Data of Robinson & Gerking, 1947.)

sweat rate to decline towards zero. Average values for the three men are shown in Fig. 7.13. The results are consistent with an exponential decline towards zero, but the time constant is about 3.8 h, considerably greater than those observed in immersion experiments. Between the first and last hours the average rectal temperature increased about 0.5 degC, so the sweating drive may have been increasing. The subjects were able to tolerate this very

7. Physiological responses

severe environment because one hand and forearm were cooled to 30.1 °C, but it is hard to see why this should affect the decline in sweating from the rest of the body surface. A possible explanation for the large time constant is that these men were also involved in experiments in which they worked in humid heat wearing poplin uniforms. Rubbing of the cloth on wet skin may have removed some of the outer epidermal layers and so rendered the sweat glands less liable to blockage.

One would not expect hidromeiosis to have a very great influence on heat balance, since until the sweat rate approaches its asymptote evaporation will not be affected. However if the stress imposed by different environments is interpreted on the basis of total sweat production over a four-hour period, as in the P4SR index, the influence of hidromeiosis might be considerable. The integrated sweating drive required to produce a given quantity of sweat during 4 h in a hot dry environment is less than that required to produce the same quantity in a humid environment where the effect of hidromeiosis is greater. Adverse effects of heat are probably better identified with the internal physiological responses, e.g. sweating drive, than with the actual quantity of sweat liberated.

Sweat gland fatigue

Sweat is hypotonic and the sweat glands must do osmotic work in order to conserve salt and other substances. It might be expected that a gland caused to work continuously for a long period might become fatigued, and that such fatigue of the gland itself, rather than of the neuroglandular junction, would be manifested as a reduction in the rate of secretion of sweat or an increase in the salinity of the sweat.

Observations by Bulmer & Forwell (1956) of the sweat rate and sweat sodium concentration during three successive half-hour periods under various conditions of environment and work are shown in Fig. 7.14. The results are plotted in accordance with the proposition that sweat is initially secreted as a plasma ultrafiltrate from which salt (and possibly water) is subsequently extracted at a rate which becomes constant when the sweat rate is high (Schwartz, Thaysen & Dole, 1953; Dobson & Lobitz, 1966). The intercept on the sodium concentration axis represents the sodium concentration in the primary ultrafiltrate, and the slope of the line indicates the rate of sodium re-absorption. Results for the first half-hour period fall around the steepest line, and for the subsequent periods the salt re-absorption is less. This appears to provide evidence of sweat gland fatigue after the first half hour, but this interpretation neglects hidromeiosis, a factor which was not recognized at the time. Sweat samples were obtained from an arm bag

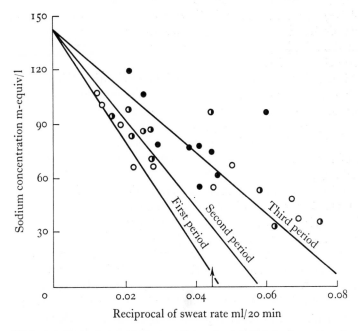

Fig. 7.14. Sweat rate and sweat sodium concentration during successive half-hour periods of work at various rates. Sweat was collected from an arm bag which was worn continuously. Samples taken during the first 10 min of each period were discarded. The abscissa shows the reciprocal of the sweat rate. At the higher sweat rates the points fall around straight lines, suggesting constant rates of sodium re-absorption from a primary isotonic fluid. The rates of sodium re-absorption diminish during successive periods. (From Bulmer & Forwell, 1956.)

which was worn continuously, and during the later collections the sweat rate must have been depressed by hidromeiosis. If some of the sweat ducts were blocked, while sweat was secreted from the unblocked glands in accordance with the line for the first period, results would have the general form of Fig. 7.14. If, for example, the sweat rate and composition from each unblocked gland were constant throughout the experiment, the sodium concentration of the collected sweat would be constant but the volume collected would decline with time, so that points for the later periods would be displaced to the right. On the basis of this interpretation the changed slopes of the lines for the second and third periods indicate the proportionate reduction in sweat rate (at any given sweating drive) due to hidromeiosis. They are consistent with an exponential decline in sweat rate towards zero with a time constant of about 110 min, in good agreement with the observations on hidromeiosis cited earlier.

This experiment emphasizes one of the difficulties of interpreting time-dependent changes in the composition of arm bag sweat. Others arise from the changes in water and salt metabolism which accompany profuse sweating and the consequent endocrine responses. It is difficult or impossible to maintain the salt concentration of the body fluids constant when large inputs and outputs are taking place, and the operation of salt or water conserving mechanisms may well affect sweat composition even if there is no sweat gland fatigue (Weiner & van Heyningen, 1952; Collins & Weiner, 1968). Sweat rate and skin temperature are known to affect the composition of sweat (Johnson, Pitts & Consolazio, 1944; Robinson, Kincaid & Rhany, 1950), and although an increase in sweat salt concentration with time during single heat exposures has frequently been observed, conclusive evidence of the existence of sweat gland fatigue is still lacking (Robinson & Robinson, 1954).

Wyndham *et al.* (1966) exposed subjects to ten different environments at five work rates. The relation between sweat rate and rectal temperature was found to change over the five-hour period. Results during the first and second hours fell on the same curve, but thereafter the sweat rate at any rectal temperature became progressively less. The finding was interpreted as evidence of sweat gland fatigue, but such fatigue was not distinguished from hidromeiosis. Curves fitted to the results for the first and fifth hours are shown in Fig. 7.15. There is no significant effect at low sweat rates, as Brown & Sargent (1965) have shown is so for hidromeiosis when evaporation is present. At higher rectal temperatures the reduction in sweat rate between the first and fifth hours is roughly proportional to the sweat rate in the first hour. This, too, is consistent with a decline due to hidromeiosis, and it cannot be concluded that sweat gland fatigue is demonstrated.

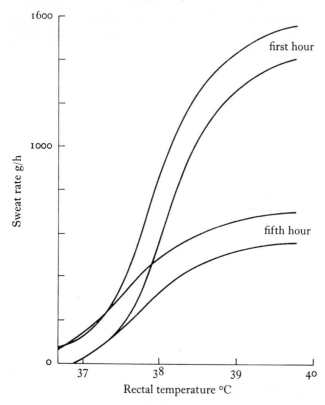

Fig. 7.15. Relation between sweat rate and rectal temperatures during the first and fifth hours of work at five rates in ten hot environments. (Redrawn from Wyndham *et al.* 1966.)

Peter & Wyndham (1966) showed by means of sweat gland counts that the number of active sweat glands diminished during exposure to a warm humid environment ($T_0 = 34$ °C, $p_a = 4.7$ kPa). This was attributed to sweat gland fatigue, any explanation based on hidromeiosis being discounted on the ground that sweating remained suppressed on the subjects' backs despite the fact that the skin did not look wet. It was pointed out earlier (p. 153) that the skin relative humidity was at least 0.8 and that the observation could be interpreted as evidence that recovery from hidromeiosis does not occur with this degree of epidermal hydration. Brown & Sargent (1965) found that sweating increased markedly when sub-

7. Physiological responses

jects were transferred from a somewhat similar environment ($T_0 = 33$ °C, $p_a = 4.0$ kPa) to a drier one, suggesting that sweat gland fatigue, if present, was not the major factor in sweat suppression.

In very long exposures to dry environments the sweat rate is usually maintained, as it must be if the body is to remain approximately in heat balance. If the sweat glands became fatigued one might expect some retention of heat which would increase the sweating drive and stimulate the fatigued glands more strongly. Such effects are not seen provided that water balance is maintained (Ferguson & Hertzman, 1958; Hertzman & Ferguson, 1959).

Thaysen & Schwartz (1955) found that the response of sweat glands to intradermal methacholine was reduced after a period of heat exposure during which the general sweat rate declined. If the glands were atropinized during the heat exposure they retained the ability to sweat later on. At the time this was interpreted as evidence of sweat gland fatigue, but later evidence suggests that hidromeiosis develops very slowly in the absence of sweat gland activity. Brown & Sargent (1965) confirmed the finding and showed further that the response of the 'fatigued' glands to methacholine could be restored by stripping the outer layers of epidermis. Collins, Sargent & Weiner (1959a) discuss the findings, pointing out that large or repeated doses of acetylcholine or methacholine may depress the sweat gland response but that the depression does not necessarily entail prolonged or intense secretory activity.

There appears to be no conclusive evidence for the existence of sweat gland fatigue as defined here. If such fatigue does occur it is probably a small factor in comparison with the effects of local skin temperature and hidromeiosis.

Blood supply

When the arterial supply to a sweating limb is occluded the sweat rate is unaffected for the first 5–15 min and declines thereafter (Kuno, 1934, pp. 118–20; Randall, Deering & Dougherty, 1948). The composition of the sweat alters during the declining phase, suggesting some impairment of gland function (Ladell, 1951; van Heyningen & Weiner, 1952b). Fig. 7.16 shows the effect of arterial occlusion on the number of active glands on the forearm (Collins, Sargent & Weiner, 1959b). Sweating was induced by heating the legs. The decline may be attributed either to impairment of sweat

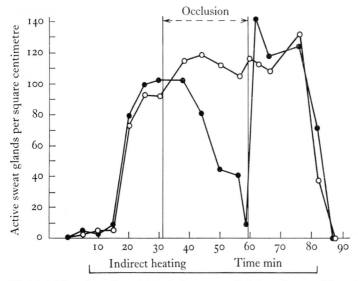

Fig. 7.16. The effect of arterial occlusion on sweating from the arm. The subject was heated by immersing the legs in warm water (indirect heating). (From Collins *et al.* 1959*b*.) ○, control arm; ●, arm circulation occluded.

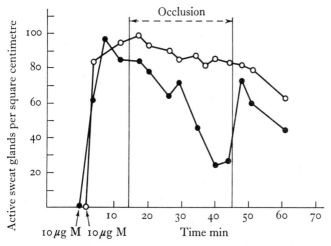

Fig. 7.17. The effect of arterial occlusion on sweating from the arm. Sweating was induced by intradermal injection of methacholine shortly before each observation. The response to this is transient, and during most of the experimental period the glands were inactive. The similarity of the effect of occlusion in this case to that in Fig. 7.16, when the glands were secreting continuously, suggests that availability of water is not a critical factor. (From Collins *et al.* 1959*b*.)

6   KSO

7. *Physiological responses*

gland function or to interference with the neural stimulating mechanism.

Fig. 7.17 shows the effect of ischaemia on sweating induced by intradermal injection of methacholine, which stimulates the glands directly (Chalmers & Keele, 1951). Large doses of methacholine depress sweating (Collins *et al.* 1959*a*) and the small doses used here produced some depression, as shown by the diminishing response of the glands of the unoccluded arm. The effect of ischaemia is clearly demonstrated. The similarity between the results shown in Figs. 7.16 and 7.17 suggests that availability of water is not a critical factor, since in the heat induced sweating the production of sweat was continuous, whereas when methacholine was used as a stimulus it was intermittent.

When the arterial occlusion is released the sweat glands rapidly recover their function (Ladell, 1951; Collins *et al.* 1959*b*). Randall, Deering & Dougherty (1948) observed a slower recovery, but sweating on the control arm was depressed when the circulation to the other arm was restored. This implies a diminution in central drive at this time, which may have obscured a rapid recovery of function in the previously occluded arm.

Acclimatization

Repeated exposure to heat stress over a period of about two weeks produces physiological changes which improve the ability of the subjects to tolerate moderate heat (Ladell, 1964). Tolerance to forced elevation of body temperature under conditions in which the physiological regulating mechanisms are ineffective is not improved (Goldman, Green & Iampietro, 1965; Williams & Wyndham, 1968), but subjects are able to regulate body temperature in more severe conditions when acclimatized than when unacclimatized. Acclimatization reduces the subjective distress produced by mild or moderate heat stress (Robinson *et al.* 1943; Eichna *et al.* 1950; Bass, 1963; Lind, 1964, pp. 16–30). If exposures to heat are discontinued, the subject reverts to the unacclimatized state in a few weeks (Bean & Eichna, 1943; Henschel, Taylor & Keys, 1943; Wyndham & Jacobs, 1957; Adam *et al.* 1960; Williams, Wyndham & Morrison, 1967).

The maximum sweat rate is greatly increased (often several times) and a given submaximal rate is achieved at lower skin or core temperature. (Belding & Hatch, 1963; Wyndham, 1967). The distribution of sweat production may change (Höfler, 1968). This improvement in sweating appears to be almost entirely due to an increase in the capacity of the glands to produce sweat. Glands in a small area of skin can be 'acclimatized' by daily local injection of stimulant drugs (Collins, Crockford & Weiner, 1965). If sweating during acclimatizing exposures is inhibited by local cooling (Fox *et al.* 1964), hidromeiosis (Brebner & Kerslake, 1963), or hyoscine (Goldsmith, Fox & Hampton, 1967), the characteristic improvement in the sweating capacity of the glands does not occur.

The control of body conductance

The thermal conductance between the body core and the skin surface can be varied by altering the skin blood flow. Like sweat production, skin blood flow is controlled by nervous impulses (constrictor and dilator) coming from the heat regulating centres. It is also affected by local skin temperature (Edholm, Fox & Macpherson, 1957; Fox & Edholm, 1963; Hellon, 1963; Thauer, 1965). Central and peripheral thermal stimuli contribute to the central vasomotor drive, and the responses to local and general heating are broadly similar to the sweat responses. It is, however, impossible to measure the total cutaneous blood flow directly, and quantitative descriptions of its control are lacking. This matters little, since it would be very difficult to calculate the relation between skin blood flow and heat transfer.

Veins commonly run close to arteries (venae comitantes), allowing ready exchange of heat between vessels (Bazett *et al.* 1949; Aschoff, 1957; Schmidt-Nielsen, 1963). The temperature drop along the length of an artery supplying blood to the skin depends on the rate at which the skin loses heat to the environment. Fig. 7.18 shows the simplest case in which blood is supposed to flow to the skin, lose heat to the environment, and return by way of a vena comitans. In passing through the skin the blood loses heat at a rate H_s. The temperature difference $(T_a - T_v)$ is H_s/F, where F is the blood flow expressed as the heat capacity of the blood passing a point in unit time. Since the flow in the vein is assumed to be equal to that in the artery this temperature difference is maintained along the length of the vessels. Assuming that the thermal resistance between the arterial and venous blood streams is entirely due to the tissues separating them, the total heat transfer between them,

6-2

7. Physiological responses

H_π, is proportional to the temperature difference $(T_a - T_v)$. The fall in temperature along the length of the artery $(T_A - T_a)$ is proportional to $(T_a - T_v)/F$, or H_s/F^2.

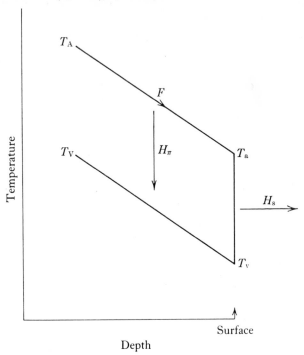

Depth

Surface

Fig. 7.18. Temperature distribution in a simple counter-current system. Heat is lost from the skin surface (H_s), lowering the temperature of the capillary blood from T_a to T_v. Blood returning along the comitans exchanges heat with blood in the artery. The flows in vein and artery are equal, so the fall in temperature along the length of the artery is equal to the rise in temperature along the length of the vein. The temperature difference between the two vessels is the same at all points. The fall of temperature along the length of the artery is not necessarily linear as in the figure.

Much of the variation in skin blood flow is due to the operation of arterio-venous anastomoses deep to the skin capillaries. When these vessels open they do not lose heat directly, but the fall in temperature along the length of the artery is reduced, so T_a is increased, raising the skin temperature and increasing heat loss. In Fig. 7.19 the total arterial blood flow divides so that F_a passes through the arterio-venous anastomosis and F_c through the capillary loops. The temperature difference between the main artery and vein at all points is $H_s/(F_c+F_a)$, and the temperature drop along the length of the artery is proportional to $H_s/(F_c+F_a)^2$. Thus for the same capillary blood flow and skin heat loss the temperature of the capillary blood supply is raised when the anastomosis opens. In practice this means that the skin heat loss increases.

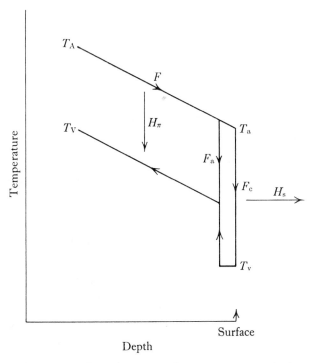

Fig. 7.19. The effect of opening an arterio-venous anastomosis deep to the capillary bed. The fall in temperature along the length of the artery is diminished because the blood flow is higher and because, for a given skin heat loss, the temperature difference between the artery and vein is less. The effect is to raise the skin temperature and increase skin heat loss.

A further important factor is the opening of superficial veins which do not run in association with arteries. Blood returning in these vessels continues to lose heat to the environment, and the skin temperature over the course of large cutaneous veins is perceptibly raised. Probably more important than this additional heat

7. Physiological responses

loss is the decrease in flow through the venae comitantes. The heat lost by the artery, $F(T_A - T_a)$ equals the heat gained by the vein, $F_{vc}(T_V - T'_V)$. Since F_{vc} is smaller than F, the temperature fall along the artery is less than it would be if all the blood returned by way

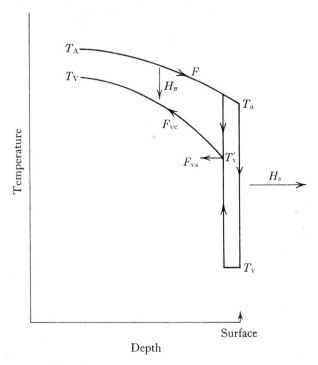

Depth

Surface

Fig. 7.20. The effect of returning some of the capillary blood through a superficial vein instead of through the vena comitans. The flow through the latter, F_{vc}, is less than that through the artery, so the temperature in this vein increases along its length more than the temperature in the artery decreases. Precooling is reduced, raising the skin temperature, and the blood returning through the superficial vein (F_{vs}) can continue to lose heat as it passes up the length of the limb.

of the vena comitans. Heat transfer between artery and vein at any point is assumed to be proportional to the temperature difference between them at that point, and is greater near the skin where the temperature difference is greater.

In all these examples the effect of blood flow on the thermal resistance between the arteries and veins has been neglected. Some

Control of body conductance

part of this resistance lies in the boundary layer of blood at the inner surface of the vessel walls, and the total resistance diminishes as the velocity of the blood stream increases.

The operation of counterflow heat exchangers (recuperators) is well understood (Jakob, 1957, Chapter 34) but cannot be usefully applied in the present context because the total blood flow, its partition, the thermal resistance between the vessel walls, and the effects of blood flow on heat transfer within the vessels are unknown.

These examples show that the relation between skin blood flow and the transfer of heat from the body core to the surface is unlikely to be simple. The complexities do not seem to trouble the physiological heat regulating mechanism, which is able to maintain a linear relation between heat loss and core temperature over a wide range of skin temperatures. Control of blood flow must be the ultimate effector mechanism, and it seems necessary for the regulating centre to be provided with information derived from the rate of heat loss. It could then adjust the blood flow to whatever level was required to provide the required rate of heat loss.

A simple mechanism which works in this way is illustrated in Fig. 7.21. The peripheral input to the controller is the temperature difference between the arterioles and venules in the skin, $(T_a - T_v)$ in Fig. 7.20 (cf. Bazett, 1951). The central input, T_c' is the difference between core temperature and a fixed reference temperature. Flow is controlled so that

$$F \propto T_c'/(T_a - T_v).$$

Since $(T_a - T_v)$ is equal to H_s/F, this may be represented

$$F \propto F . T_c'/H_s, \quad H_s \propto T_c'.$$

Such a mechanism is stable and fails only when the maximum capacity of the pump is insufficient for heat balance at the required level. The response to changes in the environment is rapid, and the flow adjusts itself to any alteration in the properties of the inter-vascular heat exchanger or to the opening of channels short-circuiting the skin blood flow provided that these are superficial to the peripheral sensor. It is not easy to see how the physiological control of blood flow could be linear, a prerequisite for this model, but a clue may lie in the dilution of vasomotor substances by the flow through the vessels upon which they act.

7. Physiological responses

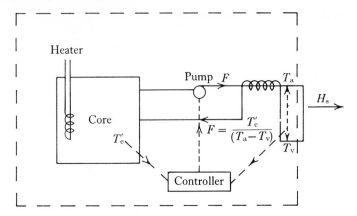

Fig. 7.21. A simple device which controls heat loss, H_s, at a level depending on T'_c, the elevation of the core temperature above a fixed reference level. A tank of water representing the body core, contains a heater (metabolic heat production). Water is pumped from the core by way of a heat exchanger to the skin surface, where heat is lost to the environment. The flow is proportional to the ratio of T'_c to the arterio-venous temperature difference, $T_a - T_v$. Flow responds rapidly to changes in the environment or in the characteristics of the internal heat exchanger. So long as the required flow is within the capacity of the pump the rate of heat loss is independent of the environmental conditions. If the rate of heat production is changed, T'_c changes exponentially to a level proportional to the new rate.

Sweating and body conductance

In view of the general similarity between sudomotor and cutaneous vasomotor responses to changes in metabolic rate or environment, it is attractive to suppose that the efferent systems are linked so that sweat production and vasodilation proceed hand in hand. There is some evidence of a peripheral mechanism for this. Sweat contains the vasodilator substance bradykinin, which, by analogy with other exocrine glands, would be expected to increase blood flow in the neighbourhood of an active gland (Fox & Hilton, 1958). Measurements of local sweat production and local blood flow suggest that the bradykinin mechanism is not very important in determining cutaneous blood flow (Senay, Christensen & Hertzman, 1961). The only available measurement indicative of total cutaneous blood flow is the body conductance, defined as $H_s/(T_c - \bar{T}_s)$. For the reasons outlined above this is unlikely to be a precise index of blood flow but it is of some interest to see how well it is correlated with sweat production.

Fig. 7.22 shows values of sweat rate and body conductance from five sets of experiments. Conductance is based on the difference between rectal temperature and mean skin temperature in all cases.

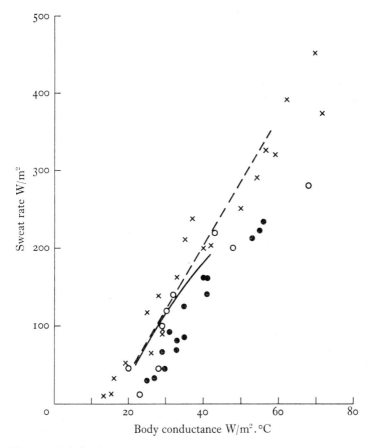

Fig. 7.22. Relation between sweat rate, expressed in equivalent heat units (W/m²) and body conductance $H_s/(T_c - \bar{T}_s)$. Data from five sources: continuous line, Nielsen, 1968; dashed line, Nielsen, 1966; ×, Wyndham *et al.* 1965*b*; ○, Stolwijk *et al.* 1968; ●, Blockley, 1965.

In view of the likely difference in heat acclimatization between the subjects the scatter of the results is surprisingly small. A wide range of work rates and environments is covered, and a linear relation between sweat rate and body conductance, regardless of work rate or environment, appears to be generally consistent with the observations. A representative equation might be

$$S = 7.5(C - 15). \tag{6}$$

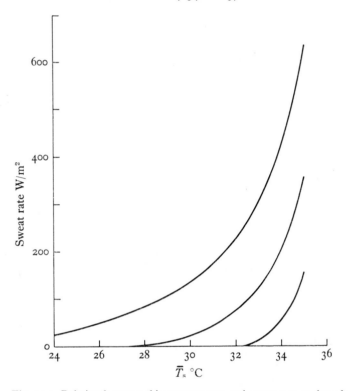

Fig. 7.23. Relation between skin temperature and sweat rate at three levels of heat production in the steady state. Based on the assumptions that $S = 7.5$ $(C - 15)$ and $T_c = 36.5 + H_s/250$. Heat loss from the respiratory tract has been ignored. The assumptions appear to be unsound, since in most subjects this relation is linear. See, however, Fig. 7.26.

It is now possible to calculate the relation between sweat rate and skin temperature at any rate of working. Within the prescriptive zone the steady state core temperature at given work rate is independent of skin temperature, and is given by equation (2). If, for example, the rate of heat production is 250 W/m², the core temperature will be about 37.5 °C. The body conductance will be $250/(37.5 - \bar{T}_s)$, and when this is substituted in equation (6) the sweat rate at any chosen skin temperature can be found. Examples are shown in Fig. 7.23.

In the next section it will be shown that direct plots of observed skin temperatures and sweat rates do not usually have the form of those in Fig. 7.23. This diagram was obtained indirectly from observations of skin temperature and sweat rate, and the distortion has appeared because a calculated core temperature has been substituted for the observed value and the sweat rate/body conductance relation has been linearized. Furthermore, values of conductance at high skin temperatures are unreliable and depend too much on the site of core temperature measurement.

Empirical description of physiological responses

The processes of human thermoregulation are not fully understood, and many mechanisms have been proposed (MacDonald & Wyndham, 1950; Hardy, 1961; Hammel *et al.* 1963; Snellen, 1966; Stolwijk & Hardy, 1966*b*; Hammel, 1965; 1968; Wyndham & Atkins, 1968; Bullard *et al.* 1970; Saltin, Gagge & Stolwijk, 1970). These are plausible concepts, and when supplied with suitable coefficients and set points may accord fairly closely with experimental observations. However since the mechanism has not been taken apart and its constituent parts demonstrated, they are, from a quantitative point of view, empirical descriptions.

In the context of the measurement of heat stress some description of the physiological reactions of the subject is necessary, and in the present state of knowledge this description must be empirical. It may as well be as simple as possible and directly related to the state of affairs at the skin surface, since this is the interface between subject and environment.

It is roughly true to say that in the prescriptive zone core temperature is independent of the environment but depends on the work rate, whereas mean skin temperature is independent of the work rate but depends on the environment (Robinson, 1949; Winslow & Gagge, 1941; Nielsen & Nielsen, 1965*a*; Snellen, 1966; Stolwijk, Saltin & Gagge, 1968; Mitchell *et al.* 1968; Nielsen, 1969; Saltin, Gagge & Stolwijk, 1970). The constancy of mean skin temperature at different rates of working is shown in Fig. 7.24 (Nielsen, 1969).

In these experiments the subject worked a cycle-type ergometer

7. *Physiological responses*

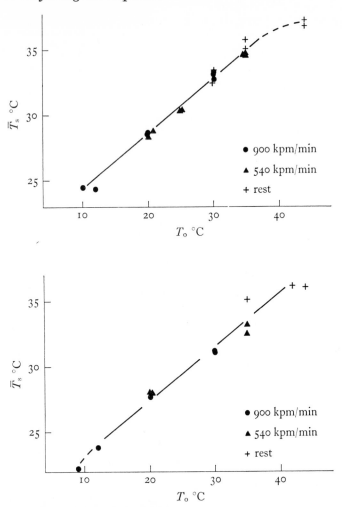

Fig. 7.24. Relation between mean skin temperature and environmental temperature for two subjects at three rates of working. (From Nielsen, 1969.)

in a reclining position. In a later series of experiments in a constant environment at 22 °C the subjects rode a bicycle mounted on a treadmill and it was found that mean skin temperature fell by about 1 degC per 100 W/m² increase in heat production. In this rather cool environment the mean skin temperatures ranged from 25 to 30 °C, and at such low temperatures there is considerable variation from place to place (Hardy & Dubois, 1938; Stolwijk *et al.* 1968). In warm environments the skin temperature is more uniform and the effect of work rate on it is less marked. There is frequently a slight fall with increasing work rate in treadmill experiments (Robinson, 1949; Macpherson, 1960, p. 139; Blockley, 1965), but it is difficult to prevent some alteration in environmental heat transfer coefficients at different rates of working. If the heat transfer coefficients differ, the environments are not identical.

If mean skin temperature is independent of work rate the requirements of heat balance demand that increments of work rate must be accompanied by exactly equivalent increases in evaporative heat loss. In a given environment if all the sweat evaporates

$$S - H_p = \text{Constant.}$$

Here H_p is the metabolic heat production, $M - W$. If the mean skin temperature is linearly related to the environmental heat load as in Fig. 7.24, the general relation for all work rates and environments in the prescriptive zone must be

$$S - H_p = K(\bar{T}_s - \bar{T}_{so}), \tag{7}$$

where $\bar{T}_{so}$ is the mean skin temperature at which sweat rate equals metabolic heat production.

Hatch (1963) adopted this general expression for sweat rate and showed that it was in good agreement with the results of Robinson (1944), Nelson *et al.* (1947) and Macpherson (1960). The agreement held for conditions of free and restricted evaporation and was not limited to the prescriptive zone. It also holds under non-steady state conditions if the skin heat loss, H_s, is substituted for H_p in equation (7) (Kerslake, 1955). If a constant thermal resistance, a, is assumed in the superficial layers of the skin the equation can be interpreted in terms of a deep skin temperature calculated as $\bar{T}_s + a \cdot H_s$. Sweat rate is then proportional to the increase in deep skin temperature above a threshold value. However it is difficult to explain why the value of K should be equal to $1/a$, as equation (7) requires, in all the series of results examined by Hatch. A more cogent objection to the deep skin temperature hypothesis is due to Nielsen (1969), who measured temperatures at various depths in the skin and showed that at no point could the measured temperature be identified with the sweat rate. No hypothesis involving core and skin temperature alone has proved able to predict the one-to-one

7. Physiological responses

relation between metabolic heat production and sweat rate in equation (7), and it seems likely that this simple empirical form is a consequence of some inter-relation between the control systems for sweating and skin heat loss.

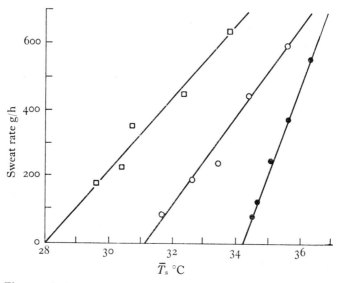

Fig. 7.25. Relation between mean skin temperature and sweat rate at three rates of working. Mean values for three subjects whose responses were similar. (From Blockley, 1965.)

Although equation (7) describes many experimental data correctly and has been used in thermal balance equations (Woodcock, Powers & Breckenridge, 1956; Hatch, 1963; Hanifan *et al.* 1963; Blockley, 1968), it does not appear to be adequate for all subjects and all conditions. Blockley (1965) found differences in the slope of the skin temperature/sweat rate curves at different rates of working. The possible differences in air movement relative to the subjects, referred to earlier, are of no consequence in this connection. The average results for three of his subjects are shown in Fig. 7.25. Both K and $\overline{T}_{s0}$ vary with the rate of working. The lines shown here have slopes which are inversely proportional to the metabolic rate. In the case of one other subject the lines were parallel, (consistent with equation (7)), and in a fifth subject, who had a thick layer of subcutaneous fat, the change in slope was considerably greater

Empirical description of physiological responses

and the lines were not straight (Fig. 7.26). This type of response cannot be described in terms of K.

Whatever the underlying mechanisms, equation (7) is a convenient description of the relation between sweat rate, skin heat loss and skin temperature for many subjects. The parameters K and $\bar{T}_{so}$ define the relation in any individual case. They are likely

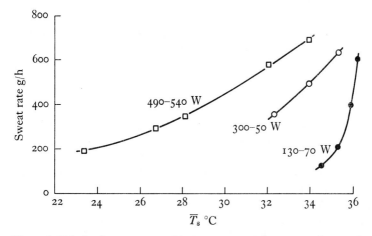

Fig. 7.26. Relation between mean skin temperature and sweat rate for one obese subject at three rates of working. (From Blockley, 1965.)

to be altered by hidromeiosis, sweat gland fatigue, heat acclimatization, dehydration and other factors affecting the performance of the sweat gland, the skin circulation or the thermoregulatory centres. Their ranges in different classes of subjects have been examined by Hanifan et al. (1963) and Blockley (1968) using original material, published figures and raw data made available to them by other investigators.

For young adult male subjects acclimatized to work in the heat, values of K range from 90 to 160 W/m². °C, and values of T_{so} from 34.6 to 33.1 °C. Acclimatization increased K and decreased T_{so}. Naturally acclimatized naval ratings in Singapore (i.e. men who had not undergone a course of severe heat exposure) had values of K ranging from 29 to 115, and unacclimatized men values as low as 15 W/m². °C. (Data from Nelson et al. 1947; Kerslake, 1955; Macpherson, 1960; Belding & Hertig, 1962; Hatch, 1963; Hall & Klemm, 1963; Hardy & Stolwijk, 1966.)

7. Physiological responses

For women the published data are less satisfactory. Observation in a single environment does not provide a value for K or T_{so} and it was necessary in some cases to assume that the sweating threshold in different groups of women would be the same. Values of K, calculated on the assumption of a linear relation between S and T_s, were in general lower than those for men, ranging from 15 to 35 W/m^2. °C. Many of the women studied were physical training students or instructresses, fitter than many of the men in the other studies. Some had been artificially acclimatized to heat. (Data from Hertig *et al.* 1963; Hertig & Sargent, 1963; Morimoto *et al.* 1967; Weinman *et al.* 1967.) Bar-Or, Lundegren & Buskirk (1969) measured the skin temperatures and sweat rates of young women working in the heat. (Skin temperature was measured during rest periods.) The observations are consistent with $K = 40$ W/m^2. °C, for both obese and lean subjects, $T_{so} = 35.3$ °C for obese, 33.6 °C for lean subjects. The data do not allow the validity of equation (7) to be checked.

The sweat rates of women at rest and in heat balance at a core temperature of 38 °C have been measured by Fox *et al.* (1969). These rates, which are probably near maximal, were much lower than for unacclimatized men. Differences in the apparent rate of hidromeiosis make direct comparison difficult, but in the first sweat collection the sweat rate was only about a third of that for unacclimatized men, which in turn was about a third of that for acclimatized men. These authors propose that the difference is essentially that women tend to be less acclimatized to heat. The inference that their sweating performance would improve dramatically with repeated exposures to work in the heat is not borne out by the observations of Hertig *et al.* (1963) and Weinman *et al.* (1967), who found that this produced little change. It can be argued that if women have an intrinsically lower sweating capacity than men they would be more frequently exposed in ordinary life to conditions demanding maximum activity from their sweat glands. If so they should be more acclimatized, though less tolerant to heat.

Wyndham, Morrison & Williams (1965) measured the sweating of male and female medical students working in the heat ($\dot{V}_{O_2} = 1.0$ l/min, $T_0 = 34$ °C, $p_a = 4.7$ kPa, $V = 0.25$–0.40 m/s). Sweat rates were greatest during the second hour, and these peak rates are shown in Table 7.1. The subjects were examined before

Empirical description of physiological responses

and after acclimatization, the sweat rates of both groups increasing in much the same ratio. The figures suggest that at the beginning the males were no more acclimatized than the females (possibly a little less) and that the females had intrinsically less sweating capacity. The acclimatization routine used in these experiments was probably more severe than those used by Hertig et al. (1963) and Weinman et al. (1967). This and the difference in local climate and culture may account for the greater response to acclimatization in Wyndham's subjects.

TABLE 7.1. *Peak sweat rates of acclimatized and unacclimatized men and women working in the heat.* (W/m². h) $\dot{V}_{O_2} = 1\ l/min$, $T_a = 34\ °C$, $p_a = 4.7$ kPa, $V = 0.25 - 0.40$ m/s. *(Data from Wyndham, Morrison & Williams, 1965.)*

	Female	Male	Ratio F/M
Unacclimatized	232	334	0.70
Acclimatized	385	602	0.64
Ratio	1.65	1.80	

There has been little work on the responses to heat of people other than young men and women. Löfstedt (1966) examined children and adults of both sexes and concluded that fit young men were able to tolerate heat stress better than older men, women of all ages or children. His investigation was directed towards establishing upper tolerable limits of heat stress for the various groups, and it seems likely that differences in sweating played a large part in determining heat tolerance. Lee & Henschel (1963, 1965) have reviewed the literature on collapse under heat stress. They do not comment on differences between men and women, but conclude that infants are less resistant to heat stress than are young adults and that tolerance to heat diminishes progressively above the age of about 45. It is lower in obese people and in those suffering from metabolic, cardiopulmonary, gastro-intestinal or skin diseases, and in the psychologically abnormal.

Lind et al. (1955) found no difference in heat tolerance of mines rescue men in the age groups 19–31 and 39–45, but men aged 41–57 were less able to stand heat (Hellon & Lind, 1958). Skin blood flow was greater in older men, suggesting that their sweating might be somewhat lower, as previous work had indicated (Hellon, Lind & Weiner, 1956; Hellon & Lind, 1956).

7. Physiological responses

The maximum oxygen consumption of males increases until about the age of 20 and declines from about 30 onwards. That of females follows the same general course but the peak is lower (about 60 per cent of the value for males) and the decline less pronounced (Astrand, 1952; Astrand, Astrand & Rodahl, 1959; Astrand, 1960; Muller, 1962). The maximum oxygen consumption of female athletes is about the same as that of sedentary men (Hermansen & Andersen, 1965). If heat tolerance or sweating capability were substituted for maximum oxygen consumption these statements would remain broadly true and the analogy is reinforced by the relation between core temperature and work rate described by Saltin & Hermansen (1966) and shown in Fig. 7.4. One cannot pursue this as far as suggesting that the maximum oxygen consumption of an individual could be used to predict his heat tolerance, since the sweating capability of sedentary people can be increased greatly by exposing them to heat at rest (Brebner et al. 1961; Fox et al. 1963), a procedure which probably has little effect on their maximum oxygen consumption.

8 EQUIVALENT ENVIRONMENTS–STEADY STATE

The factors governing heat exchange between the skin and environment have been examined in the preceding chapters and can be classified as follows:

Environment
  Air temperature
  Mean radiant temperature (infra-red)
  Sunlight
  Water vapour pressure
  Air movement
  Atmospheric pressure

Clothing
  Insulation
  Wind penetration (related to ambient air movement)
  Ventilation (related to ambient air movement and movement of the subject)
  Water vapour permeability, wicking characteristics
  Transmittance (radiation)
  Emittance (infra-red)
  Absorptance (sunlight)

Physical properties of subject
  Shape, size and physical movement (related to ambient air movement)
  Radiant area
  Projected area (direct sunlight)
  Emittance of skin (infra-red)
  Absorptance of skin (sunlight)

Properties of the subject which depend on his physiological responses
  Skin temperature
  Sweat rate

8. Equivalent environments – steady state

Respiratory exchange
Insensible water loss.

It would be useful if these factors could be combined so that the effects of different environments, work rates and clothing could be compared. Preferably a single figure would describe the thermal situation. This chapter examines the feasibility of basing an index of steady heat stress on physical principles.

Stress and strain

A physical object, such as a spring, can be subjected to a stress imposed from outside, in this case a force or load, which will induce a strain in the object, in this case a change of length. The stress can be described without reference to the strain it produces, and can be regarded as a pure component of the environment. The strain it induces in different springs depends on the properties, in this case elastic moduli, of the springs.

It would be convenient if a similar distinction could be drawn in the case of human heat stress, the stress to be defined solely in terms of the environment, the strain to be a consequence of the stress and the physiological characteristics of the subject. Stress and strain must be defined in such a way that they are distinct but can be related through the subject's responses. Hatch (1963) considers that a satisfactory scale of heat stress must specify the stress (imposed by environment and clothing) and strain (physiological state) independently. An expression of heat stress which fully met this requirement would be applicable to both steady and non-steady states of the subject and to all rates of working. For the present it will be supposed that the strain is related only to the physiological state of the subject (e.g. sweat rate, skin blood flow). Subjective judgements unrelated to physiological responses will be ignored.

If heat stress can indeed be expressed independently of physiological responses, and so be defined purely in terms of the environment and clothing, it follows that families of equivalent environments (i.e. imposing the same heat stress) must be the same for all subjects. The strain induced in subject A by heat stress X may be different from that induced in subject B by the same stress, but all environments inducing this strain in subject A must also induce the equivalent strain in subject B.

If the environmental heat stress is to be expressed by a single number, Hatch's requirement cannot be met. Suppose subject A does not sweat at all. His heat exchange must be independent of humidity, and for this subject the families of equivalent environments at constant air movement will be those of equal Operative temperature. If subject B sweats so much that his skin is always wet, his families of equivalent environments will be wet surface isotherms and each will cover a wide range of Operative temperatures. They will therefore differ from those of subject A, and the attractive analogy of stress and strain will not work.

If the metabolic heat production and all but one of the environmental components are constant, the remaining component will behave as a pure environmental stress, and can be expressed by a single number irrespective of the nature of the subject. If, for example, humidity were the only variable it would not matter whether this stress was expressed as vapour pressure, dew point, wet bulb temperature or any of the other expressions used for humidity. Subject A, who does not sweat, would have a physiological response curve in which the strain was independent of this stress, whereas the strain in subject B would vary with the stress, the form of the curve depending on the method chosen for expressing humidity.

It is no use treating each environmental component as a separate stress entity, since this would limit equivalent environments to those which are identical in every respect. One must seek to combine the components in some way even if the whole environmental complex cannot be reduced to a single number. The chief interface between the subject and environment is the skin. (For the present the small heat exchange in the respiratory tract will be ignored.) Heat is exchanged at the skin surface either as sensible heat, when the rate depends on temperature differences, or as latent heat, when it depends on sweat rate, possibly limited by vapour pressure difference. The subject's physiological responses can alter the rates of sensible and latent heat transfer, often independently of one another, but cannot alter the partition of sensible heat exchange between radiation and convection. As a first step it is worth examining how well the environmental components which affect sensible heat exchange can be combined and expressed as a single measure of sensible heat stress for a nude subject.

8. Equivalent environments – steady state

Operative temperature

If heat exchange by conduction (contact with solid objects) is neglected, sensible heat transfer takes place only by convection and radiation. The three factors concerned, air temperature, radiant temperature and air movement, can be reduced to two by using the Operative temperature approach (p. 66). However in doing this certain properties of the subject must be introduced (Herrington, Winslow & Gagge, 1937). These are not physiological responses but the physical characteristics of size, shape, emittance and orientation, which interact with the environmental components. In the conventional exposition of Operative temperature the environment is described by the air and radiant temperatures and the coefficients, h_c and h_r. The convection coefficient, h_c, depends on the air movement and the size, shape and orientation of the subject. The first power radiation coefficient, h_r, depends on the radiation area, a function of the shape of the subject, and the emittance factor, a function of the physical properties of both the subject and the radiant enclosure. (The dependence of h_r on temperature is not directly relevant here, and can be side-stepped by using the effective radiant temperature (p. 58).)

Thus the most elementary step in combining environmental components involves some physical properties of the subject, but apart from voluntary changes in posture these are not under physiological control. If we are prepared to accept this degree of intrusion of subject characteristics into the formulation of heat stress (as opposed to strain), the next step is to enquire to what extent environments of equal atmospheric pressure, air movement, humidity and Operative temperature can be considered equivalent. It will be assumed that the Operative temperature is calculated appropriately for the subject's physical characteristics. This calculation is based on the mean heat exchange coefficients, $\bar{h}_c$ and $\bar{h}_r$, for the whole body, and leads to families of environments in which air temperature and mean radiant temperature differ but the mean sensible heat exchange (regardless of skin temperature) is the same. If the local values of h_c and h_r were the same at all points on the skin surface environments of equal Operative temperature would be truly equivalent, but in practice these coefficients and their ratio vary from place to place. The weighting of air and radiant temperatures appropriate for one skin region is not necessarily appropriate for another, and it is only at places where the local ratio h_c/h_r happens to equal the ratio of the mean coefficients, $\bar{h}_c/\bar{h}_r$, that sensible heat exchange in differently constituted environments of the same Operative temperature will be constant.

In the same way as Operative temperature is calculated from the mean coefficients, a local Operative temperature can be found from the local coefficients. In the case of a cylinder in a transverse wind the smallest value of h_c is about one third of the greatest (upstream position) (Fig. 2.7). Suppose that in a particular case the actual values are 3.0 and 9.0 W/m². °C. The mean value for the whole cylinder is about 6.0 W/m². °C (from Fig. 2.7, and only by coincidence the arithmetic mean of the maximum and minimum). Assume that h_r is 6.0 W/m². °C at all points. The (mean) Operative temperature is given by

$$T_0 = \frac{\bar{h}_c}{(\bar{h}_c + \bar{h}_r)} . T_a + \frac{\bar{h}_r}{(\bar{h}_c - \bar{h}_r)} . \bar{T}_r = 0.5 T_a + 0.5 \bar{T}_r.$$

For the upstream position, where h_c is 9.0, the appropriate weighting would be $0.6 T_a + 0.4 T_r$, and at the flank where h_c is 3.0 the weighting should be $0.33 T_a + 0.67 T_r$. If $T_a = T_r = 20$ °C the weightings are immaterial and the local Operative temperatures are all 20 °C. However, if $T_a = 10$ °C, $T_r = 30$ °C the mean Operative temperature is still 20 °C, but the local Operative temperatures at these two sites are 18 and 23.3 °C. In this condition the well-ventilated upstream position will be effectively in a cooler environment than the poorly ventilated flank. Differences of the same sort and order are to be expected in the case of a human subject, even in an omnidirectional air movement, since h_r varies from place to place as a result of opposition of different skin surfaces. In no two differently constituted environments of equal Operative temperature will the distribution of skin heat loss be the same, and while subjects may judge the general level of heat stress to be equal, the sensations produced cannot be identical.

Subjective comparison of the sensations produced by environments of moderate or severe heat stress is difficult and unreliable, but judgement of situations at or near comfort can be made with some assurance. Fig. 8.1 shows the results of an experiment in which a nude resting subject was asked to adjust the intensity of radiant heating so as to maintain thermal neutrality or a state of comfort (Gagge, Stolwijk & Hardy 1965). The air movement was

8. *Equivalent environments – steady state*

low (less than 0.07 m/s) and one would expect the Operative heat transfer coefficient, h_0, to be about 7.0 W/m². °C (Stolwijk, Hardy & Rawson, 1962; Gagge & Hardy, 1967). If the Operative temperature selected by the subject were constant, the points would fall

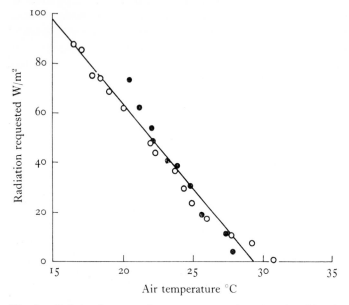

Fig. 8.1. Relation between air temperature and preferred radition intensity. Subjects were allowed to select the radiation intensity, here expressed as the radiant heat gain, at various air temperatures in an environment of low air movement. Open symbols show results obtained when the previous air temperature was lower, filled symbols when it was higher. If the Operative temperature selected by the subject is constant the line should have a slope of $-h_0$W/m². °C. The intercept on the temperature axis is the preferred Operative temperature.

on a straight line with slope $-h_0$. (In terms of the Effective Radiant Field (p. 73), $T_0 = T_a + H_r/h_0$.) The slope of the line shown in Fig. 8.1 is -6.8 W/m². °C, close to the expected value, and similar results were reported on other subjects. At low air temperatures the subjects were able to select radiation intensities consistent with the 'neutral' state on a seven point thermal vote scale, but could not always identify the state as one of comfort. This doubtless reflects differences in skin temperature distribution, but nevertheless it appears that the general level of thermal sensation was closely related to the Operative temperature.

Standard Operative temperature

Standard Operative temperature

Since Operative temperature is successful in combining air and
radiant temperatures when air movement is constant, it is worth
seeing whether the same criterion of equality of sensible heat
exchange can be applied to cases in which the air movement is
different. In his original proposal for a scale of 'Standard Operative
temperature', Gagge (1940; 1941) used equations containing air
velocity rather than the consequent convective heat exchange coefficient. The resulting expressions are rather cumbersome, and
more recently Gagge & Hardy (1967) have suggested the simpler
approach of using the Operative heat exchange coefficient.

A standard reference environment of low air movement (0.07 m/s)
is imagined in which air and radiant temperatures are equal. This
temperature is called the Standard Operative temperature, T'_0. The
combined coefficient for convection and radiation for this environment, h'_0, has the value 7.0 W/m². °C. Sensible heat exchange in
this reference environment is given by

$$(C+R) = h'_0 (T_s - T'_0). \tag{1}$$

In any other environment the sensible heat exchange is given by
a similar equation containing the Operative temperature, T_0, and
Operative heat transfer coefficient, h_0, for that environment.

$$(C+R) = h_0(T_s - T_0). \tag{2}$$

The Standard Operative temperature at which sensible heat
exchange would be the same as it is in some actual environment T_0,
h_0, can be found by equating the right-hand sides of equations (1)
and (2).

$$T'_0 = \frac{h_0}{h'_0}.T_0 - \left(\frac{h_0}{h'_0} - 1\right) T_s. \tag{3}$$

For the case $h_0 = h'_0$ (i.e. the air movement in the actual environment is the same as in the reference environment) the skin
temperature term is eliminated, but in all other cases it remains.
Environments of different air movement can only be reduced to
equivalent terms if the skin temperature is known.

The reason why this must be so can be seen by considering two
environments of high air movement in which T_0 is respectively
higher and lower than T_s. When T_0 is higher than T_s the temperature of the reference environment (T'_0, h'_0) would have to be higher

8. Equivalent environments – steady state

than T_0 in order to achieve the same gain of sensible heat at this lower air movement; when it is lower, T'_0 would have to be lower than T_0 to achieve the same heat loss. Thus the relation between T_0 and T'_0 must depend on skin temperature as well as on air movement.

As in the simpler case of Operative temperature, differences in the distribution of heat exchange over the skin surface are to be expected in differently constituted environments of equal Standard Operative temperature, but since the criterion of equality of mean sensible heat exchange works satisfactorily for Operative temperature it is unlikely that these differences will constitute an important objection in practice.

The crucial problem is that as soon as air movement is brought into the equations, skin temperature comes in too, and this is affected by the subject's responses. Standard Operative temperature is a compound of both the environmental heat stress and the strain on the subject, and so fails to meet Hatch's requirement for a measure of stress which is independent of strain. If we wish to express the sensible heat stress by a single number (T'_0) we must abandon the distinction between stress and strain and allow the value of T'_0 (for any environment in which h_0 is not equal to h'_0) to depend on physiological characteristics of the subject. The Standard Operative temperature of a given environment is different for different subjects.

Alternatively, the distinction between stress and strain can be retained if more than one number is used to describe the stress. One could, for example, use the line relating T'_0 to T_s (see equation (3)), thus removing the need to define a particular skin temperature, and in effect leaving the subject the option of responding as he pleases. However two numbers (the slope and position of the line relating T'_0 and T_s) are then required in order to describe the environment, and these replace the numbers, T_0 and h_0, which previously described it. This is not encouraging, but it is worth proceeding further along these general lines.

Equivalence based on total heat exchange

Environments of low humidity in which mean sensible heat exchange is the same are sufficiently similar in their effects to allow them to be described as equivalent (Fig. 8.1). Possibly this criterion of equivalence could be extended to cover environments in which the total heat exchange (sensible plus latent) is the same. The rate of evaporative heat loss required to produce a total skin heat loss H_s (in the steady state $H_s = (M - W)$) is given by

$$E_{req} = H_s - h_0(T_s - T_0).\qquad(4)$$

There is a maximum rate of evaporation, E_{max}, which cannot be exceeded, but below this limit the subject can achieve the specified rate of heat loss, H_s, at various skin temperatures and rates of evaporation.

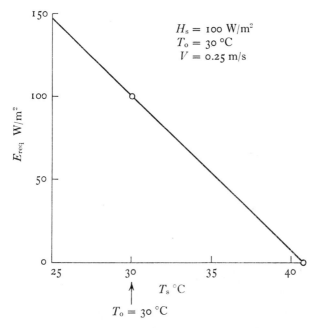

Fig. 8.2. Required rate of evaporative heat loss in an environment of Operative temperature 30 °C, air movement 0.25 m/s. The rate of heat loss from the skin is 100 W/m².

Figure 8.2 shows the relation between E_{req} and T_s for a skin heat loss of 100 W/m², Operative temperature 30 °C, air movement 0.25 m/s. The line intersects the temperature axis at $T_s = T_0 + H_s/h_0$. At $T_s = T_0$, $E_{req} = H_s$. These two points enable the line to be plotted easily. Its slope is $-h_0$, and its position depends on T_0 and H_s. It has been necessary to combine metabolic

8. *Equivalent environments – steady state*

heat production with environmental factors in order to produce the diagram, but it is not unreasonable that the rate which at heat must be transferred should appear in the statement of heat stress.

We must now turn to the question of limitation of evaporation. The lower the skin temperature, the greater the required rate of

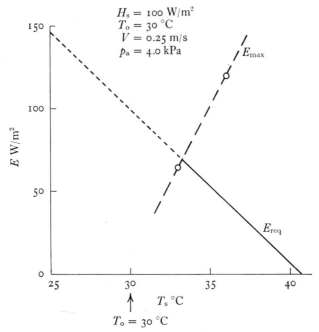

$$H_s = 100 \ \text{W/m}^2$$
$$T_0 = 30 \ °\text{C}$$
$$V = 0.25 \ \text{m/s}$$
$$p_a = 4.0 \ \text{kPa}$$

$T_0 = 30 \ °\text{C}$

Fig. 8.3. Limitation of evaporation. The line E_{req} is the same as that in Fig. 8.2 and relates to the same conditions. The line E_{max} shows the maximum evaporative capacity at an ambient water vapour pressure of 4.0 kPa. The intersection of these lines defines the lowest skin temperature consistent with the required rate of heat loss (100 W/m²). Below this the required rate of evaporation is greater than the maximum evaporative capacity.

evaporation and the lower the vapour pressure difference available to produce it. The condition $E_{req} = E_{max}$ therefore occurs at the lowest skin temperature consistent with heat balance. The position of this point can be established by the construction shown in Fig. 8.3. Two skin temperatures are chosen, say, 33 and 36 °C ($p_s = 5.03$ and 5.94 kPa). E_{max} is found for each of these temperatures from the equation $E_{max} = h_e(p_s - p_a)$. The intersection of the line through these points with the line for E_{req} defines the point

$E_{req} = E_{max}$. That part of the E_{req} line to the left of this point is unreal because evaporation could not take place at the required rate. Beyond the intersection with the temperature axis the line is also unreal because the rate of evaporation cannot be negative. (Condensation cannot occur if evaporation is possible at lower skin temperatures.)

The continuous line in Fig. 8.3 depends on the three environmental factors, Operative temperature, air movement and humidity. In order to define the line one could use, for example, the value of T_s at $E_{req} = 0$ and the values of T_s and E_{req} at $E_{req} = E_{max}$. Alternatively one could use the first point, the slope and the length of the line, but whatever method is used three numbers must be involved. The three environmental factors have been replaced by the three parameters required to define the stress line, and no economy of expression has been achieved. Any advantage in this form of presentation must lie in the dimensions of the quantities used, which may bear a closer relation to the physiological responses of the subject than do the original environmental measurements.

The stress and strain relation

In order to apply the concept of stress and strain we must be able to define the relation between them for a particular subject. This relation, which is descriptive of the subjects' responses, must be expressed in terms which can be applied both to the environment and to the physiological strain in the subject.

The transfer of heat from the deep tissues to the skin surface depends on the core temperature (which need not be closely defined here), the skin temperature, and the effective conductance between the deep tissues and the skin, a function of the skin circulation. Skin temperature is influenced by the environment, but the rate of heat transfer from the deep tissues to the skin at given skin temperature is not. Skin temperature is therefore a quantity directly and separately involved in heat transfer inside and outside the body, and may properly constitute one of the terms in which both stress and strain may be expressed.

Evaporation is not under direct physiological control. The relevant physiological process is sweat secretion, and the evaporative heat loss which results is determined by environmental factors. Evaporation cannot be used in the expression of strain, but if one could express stress in terms of required sweat rate rather than

8. *Equivalent environments – steady state*

required rate of evaporation, the expressions for stress and strain would be compatible.

If the rather small differences in the total heat of evaporation, which depend on the relative humidity of the skin (p. 27) are ignored, the ratio of evaporative heat loss to sweat rate, η_s, is a function of the wettedness, W (p. 46). The wettedness required for a particular rate of evaporative heat loss and skin temperature is determined by environmental factors. It is therefore appropriate that the relation between W and η_s should appear in the stress rather than in the strain formulation. This relation depends on the distribution of local values of h_e and also on the distribution of sweat rate. The former is a consequence of the shape of the subject, already appearing in the stress formulation, and of the nature of the air movement, also indirectly present there. The distribution of sweat production, on the other hand, even if it were independent of total sweat rate, should not appear in the stress formulation, since it is part of the subject's response. Unfortunately the pattern of sweat distribution cannot be separated from the other factors determining η_s, so it is impossible to express stress in terms compatible with strain without involving some of the physiological responses of the subject.

At this point the simplest course would be to abandon any attempt at rational expression of heat stress, and to construct an index of strain based on empirical observations. Unfortunately this would make it impossible to predict the effects on subjects other than those on whom the observations had been made. It might well be preferable to accept a rationally based expression of heat stress, even if this were imperfect, provided that the more important subject responses were not involved. In accepting Operative temperature as a measure of a sensible heat stress it was necessary to concede that physical properties of the subject should appear in the stress formulation and that differences in the distribution of skin heat loss should be ignored. Differences in sweat distribution (not total sweat rate) between different subjects and at different sweat rates (Weiner, 1945; Ferguson *et al.* 1956) may not be great enough to invalidate the expression of heat stress in terms of skin temperature and required sweat rate, and such a mode of expression would allow both the heat stress and the subject's responses to be stated in the compatible form of skin temperature/sweat rate relations.

Equivalence based on total heat exchange

Some limits can be set to the probable relation between wetted area and sweating efficiency for nude subjects. It is unlikely that the distribution of sweat production will be such that regions which are normally poorly ventilated will sweat more than the well ventilated regions. (Sweating from the axilla is an obvious exception, but the glands here are not primarily thermal in function, and can be stimulated by chilling the subject (Glaser & Lee, 1953).) The

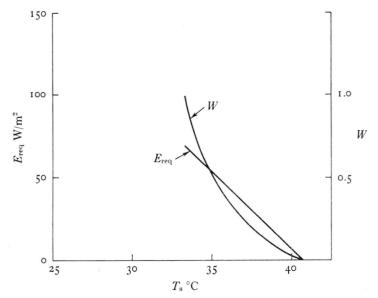

Fig. 8.4. Relation between skin temperature and wettedness for the conditions of Fig. 8.3 ($T_o = 30$ °C, $p_a = 4.0$ kPa, $V = 0.25$ m/s, $H_s = 100$ W/m²). The wettedness, E_{req}/E_{max}, is unity when the skin is completely wet with sweat and zero when there is no evaporation. The line for E_{req} is the same as that in Fig. 8.3.

subject may therefore be expected to show a somewhat higher sweating efficiency than if he were to sweat uniformly all over. Unacclimatized subjects often sweat excessively from the face, but as the head is usually at the top of the body the excess sweat runs down on to the trunk and may evaporate from there. The greatest local variations in evaporation coefficient are to be expected (for the nude subject) in a unidirectional wind, and other forms of air movement are likely to provide more uniform ventilation of the skin. The example of the uniformly sweating cylinder in a wind may therefore represent the worst case. Sweating efficiency in nude

8. *Equivalent environments – steady state*

subjects is likely to be higher, and in some circumstances may approach the ideal of no wastage until maximum evaporation is reached (cf. Fig. 2.6).

Figure 8.4 shows the line for E_{req} from Fig. 8.3 and also the subject's wettedness, W, which can be calculated without reference to sweating efficiency since it is merely E_{req}/E_{max}. It is now possible

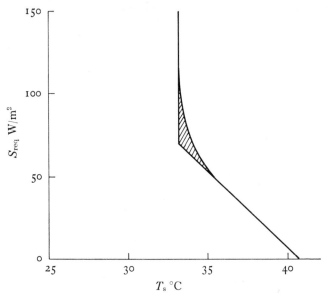

Fig. 8.5. Relation between skin temperature and sweat rate for the conditions of Fig. 8.4. The two straight lines show the relation for the ideal case in which there is no wastage of sweat by dripping until the skin becomes completely wet at all points. The curved line shows the relation for the case of a uniformly sweating cylinder in a transverse wind. The shaded area shows the zone within which the curve for a nude human subject probably lies.

to find the relation between skin temperature and required sweat rate. This is shown in Fig. 8.5. Two cases are represented. In the first the posture and sweating distribution are ideal and S_{req} is the same as E_{req} until full wetness is reached. Beyond this point S_{req} becomes infinite. The other case is that of the uniformly sweating cylinder in a unidirectional wind. Here the required sweat rate is E_{req}/η_s, η_s being found by entering Fig. 2.10 at the appropriate value of W. The shaded area in Fig. 8.5 marks the difference between the two cases, and the curve for a nude subject probably lies within this region.

Operating conditions

At a given rate of heat production in the steady state the subject's sweat rate is related to his skin temperature by a curve which is characteristic for that subject (Chapter 7). This curve defines the subject's physiological response. The complementary curve defining the environmental stress is that of Fig. 8.5 which shows the sweat rate necessary to achieve heat balance for a specified environment and rate of heat production. The skin temperature and sweat rate at which a particular subject will operate under these conditions are defined by the intersection of the stress and response curves.

Response curves for two hypothetical subjects are shown in Fig. 8.6. Subject A is a young man, moderately acclimatized to heat. At the specified rate of heat production (100 W/m²) he begins to sweat at a skin temperature of 33.3 °C and increases his sweat production by 100 W/m² per degree rise in skin temperature. Subject B is a young woman having a sweating threshold of 34 °C and an increase of only 15 W/m².°C (Blockley, 1968).

The operating conditions of the two subjects are very different. The skin temperature of subject A is between 34.1 and 34.2 °C, sweat rate between 62 and 70 W/m², depending on the value of η_s. His wettedness is about 0.7. Subject B has a skin temperature of 36.6 °C, sweat rate 38 W/m², and her wettedness is 0.3. This state of affairs sustains the convention that men perspire while ladies merely glow. However the problem of transferring body heat to the skin is very much greater for subject B. At a core temperature of 37.2 °C, which would be reasonable for this rate of heat production in the prescriptive zone, she would require a mean body conductance of 167 W/m².°C, which is unattainable (cf. Fig. 7.22). Her core temperature would therefore rise until the difference between core and skin temperatures was large enough for heat balance to be attained. She would be operating beyond her prescriptive zone and her internal heat transfer mechanisms would be fully deployed. Subject A has no comparable problem. At a core temperature of 37.2 °C his core–skin temperature difference would be about 3 °C requiring a body conductance of only 33 W/m².°C. This circulatory demand is easily met. Thus A is wet and happy while B is hot and miserable.

8. *Equivalent environments – steady state*

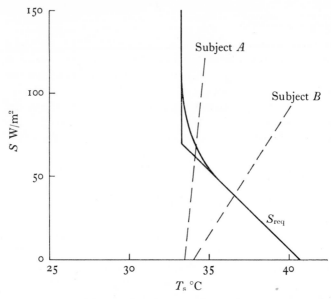

Fig. 8.6. Operating conditions for two subjects. The continuous lines show the sweat rate required for heat balance, and are identical with those in Fig. 8.5. The dashed lines show the relations between skin temperature and sweat rate for two subjects. Subject A is a young man, moderately acclimatized to heat, subject B an unacclimatized young woman. The points where the response curves intersect the line for S_{req} indicate the skin temperatures and sweat rates at which the subjects will be in heat balance.

Equivalent environments

The operating condition of any subject can be found in this way, and it would be reasonable to define equivalent environments as those in which the operating conditions are the same. However a certain combination of sweat rate and skin temperature in a given subject is consistent with only one rate of heat production, so this proposal must be limited to cases in which the rate of heat production is the same. Accepting this restriction, equivalent environments at a given rate of heat production are those in which the skin temperature is the same. The sweat rate for a given subject is also the same, since it is defined by his response curve. The families of equivalent environments may be constructed as skin isotherms on the psychrometric axes.

The construction of skin isotherms for conditions of constant skin heat loss and sweat rate was described in Chapter 4. The

Equivalence based on total heat exchange

present criterion is overall heat balance rather than constant skin heat loss, and the lines must be modified to allow for respiratory heat loss. It is assumed that at constant skin temperature and heat production the mass rate of sweat production will be constant. Diffusional water loss is additional to this, and the total heat of evaporation, which depends on the skin relative humidity, must be applied to all the skin water loss.

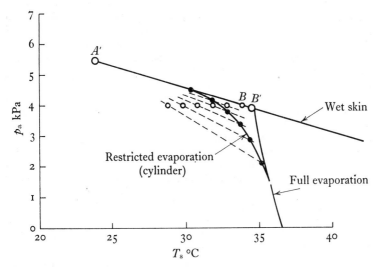

Fig. 8.7. Construction of a skin isotherm, $T_s = 34.5$, for a nude subject in the steady state producing heat at the rate of 100 W/m². Wind speed 0.25 m/s, sweat rate (expressed as latent heat of vaporization) 100 W/m², respiratory minute volume 18 l/min. For explanation see text.

The isotherm in Fig. 8.7 is appropriate for a subject having a heat production, H_p, of 100 W/m², skin temperature 34.5 °C, sweat rate (expressed as latent heat of vaporization, not total heat of evaporation) 100 W/m², and respiratory exchange 18 l/min. The wind speed is 0.25 m/s, h_o 9.3 W/m². °C, h_e 62 W/m².kPa. The wet skin line is based on the points A' and B'. At the point A', p_a is equal to p_s (in this case 5.47 kPa), so there is no evaporative heat loss from the skin. The approximate value of T_o is found on the assumption that all the heat produced is lost by sensible heat transfer from the skin surface; $T_o = T_s - H_p/h_o$. In this case $T_o = 23.75$ °C. The respiratory heat loss in this environment, found from Fig. 6.2, is 9 J/l. This is multiplied by the respiratory exchange (expressed in l/m².s) to give the respiratory heat loss in W/m², in this case 1.5. The skin heat loss required for thermal balance is therefore not 100 W/m² but 98.5 W/m², and the true value of T_o at point A' is 23.91 °C. The value used in entering

7-2

8. Equivalent environments – steady state

Fig. 6.2 was not quite correct, but reference to that figure shows that the resulting error in the estimate of skin heat loss is negligible.

Point B' corresponds to the condition of no sensible heat exchange, $T_o = T_s$. The approximate value of p_a, on the assumption that all the heat is lost by evaporation from the (wet) skin surface is $p_s - H_p/h_e$, in this case 3.86 kPa. For this environment the respiratory heat loss is 5.3 W/m². The true heat loss from the skin is therefore 94.7 W/m² and the correct value of p_a is 3.94 kPa. The wet skin line is constructed by joining the points A' and B'.

The line for full evaporation is constructed as follows. The approximate value of T_o is $T_s + (S - H_p)/h_o$, in this case 34.5 °C. A value for p_a is now chosen, say 1.0 kPa. The respiratory heat loss is 12 W/m². The next step is to calculate the approximate skin relative humidity from the equation, $\phi_s = (p_a + S/h_e)/p_s$. In this case $\phi_s = 0.48$. The diffusional water loss through the skin, expressed as heat of vaporization, is $1.2\, p_s(1 - \phi_s)$, in this case 3.4 W/m². This is added to the sweat rate to give the total skin water loss expressed as heat of vaporization, in this case 103.4 W/m². The heat removed from the skin by evaporation is greater than this. Fig. 2.2 shows that at $\phi_s = 0.48$, the ratio of total heat of evaporation to heat of vaporization is 1.042. The evaporative heat loss from the skin is therefore 107.7 W/m². The metabolic heat to be lost through the skin is equal to the heat production less the respiratory heat loss, in this case 88.0 W/m². Skin evaporative heat loss exceeds this by 19.7 W/m², so the Operative temperature must be $19.7/h_o$, or 2.12 degC above skin temperature. The point on the full evaporation line is therefore 36.12 °C, 1.0 kPa. The procedure is repeated at other values of p_a.

The method is somewhat approximate, but it is only at very low sweat rates and ambient water vapour pressures that the errors are important. For example at $S = 0$, $p_a = 0$, the diffusional water loss determines ϕ_s and cannot be neglected in computing the approximate value of ϕ_s.

The two lines constructed in this way provide the isotherm for the ideal case in which the transition from full evaporation to full wetness occurs at the same ambient conditions in all skin areas. The line for the probable worst case, the uniformly sweating cylinder in a wind, can be constructed from this as described on p. 92, point B being chosen 10 degC below point A' and on the wet skin line.

Figure 8.8 shows isotherms for a subject in the steady state who sweats according to the equation.

$$S = H_p + 100(T_s - 34.0).$$

The heat production, H_p, is 100 W/m² and the air movement 0.25 m/s. The lines for full evaporation are broadly similar in shape and slope despite large differences in sweat rate (from zero to 200 W/m²). One might expect the effect of increasing heat of evaporation at low humidities to be more evident at high sweat rates. However when the sweat rate is high the skin relative humidity is raised, and at high levels of skin relative humidity the effect of changes in ambient humidity on heat of evaporation is small (cf. Fig. 2.2). It happens that this roughly cancels the increased sweat rate. The ideal isotherms (lines for full wetness and full evaporation) can be superimposed fairly closely by shifting the axes, but this is not true of the isotherms for the cylinder, which cut the corners of the ideal isotherms. These are geometrically similar in shape but differ in size.

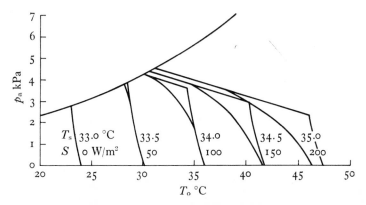

Fig. 8.8. Skin isotherms for a heat production of 100 W/m², wind speed 0.25 m/s. The subject sweats according to the equation $S = H_p + 100(T_s - 34.0)$. At each skin temperature curves are shown for the ideal case and for the case of a uniformly sweating cylinder.

Fig. 8.9 shows isotherms for the same conditions except that the air movement is 1.0 m/s. The wet skin lines slope less than those in Fig. 8.8, and the isotherms are closer together on the temperature axis, reflecting the difference in h_o. The generally higher level of the wet skin lines at the higher wind speed is a consequence of the higher value of h_e, as is the smaller difference between the cylinder and ideal curves. Isotherms from the two figures cannot be superimposed.

Fig. 8.10 shows isotherms for a subject who sweats according to the equation
$$S = H_p + 50(T_s - 34.0).$$

The air movement is 1.0 m/s as in Fig. 8.9. The isotherms for $T_s = 34.0$ in Fig. 8.9 and 8.10 are identical, since in each case the

8. Equivalent environments – steady state

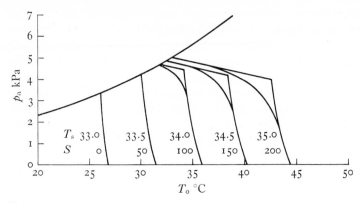

Fig. 8.9. Skin isotherms for the same subject as Fig. 8.8. Heat production 100 W/m², wind speed 1.0 m/s.

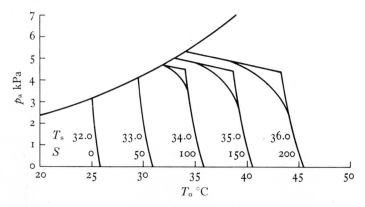

Fig. 8.10. Skin isotherms for a heat production of 100 W/m², wind speed 1.0 m/s. The subject sweats according to the equation $S = H_p + 50(T_s - 34.0)$.

sweat rate is 100 W/m². The other isotherms have obvious similarities and those in which the sweat rate is the same can be nearly exactly superimposed by shifting the axes a little. At $S = 200$, the isotherm in Fig. 8.9 is for $T_s = 35.0$, $p_s = 5.62$ kPa. That in Fig. 8.10 is for $T_s = 36.0$, $p_s = 5.94$ kPa. Fig. 8.11 shows how closely these two curves correspond if the axes are adjusted by the difference in skin temperature and in skin vapour pressure. The lines for wet skin and restricted evaporation are almost identical. Those for full evaporation differ by about 0.2 degC on the temperature axis.

This suggests that lines of equal sweat rate (isohids) for different subjects have more in common than do isotherms, and might be a more convenient way of expressing environments of equal heat stress. A separate set of curves is required for each air movement (cf. Figs. 8.8 and 8.9), but at a given air movement a single set of isohids can easily be adjusted for different subject characteristics.

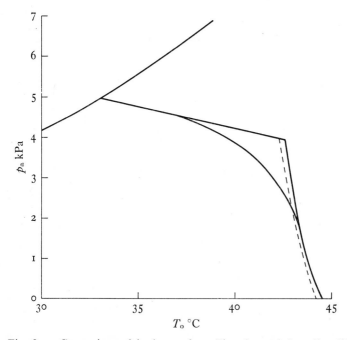

Fig. 8.11. Comparison of isotherms from Figs. 8.9 and 8.10 (for different subjects), heat production 100 W/m², wind speed 1.0 m/s. Both curves are for a sweat rate of 200 W/m². The continuous lines are reproduced from Fig. 8.10, $T_s = 36.0$ °C, $p_s = 5.94$ kPa. The dashed lines show the isotherm from Fig. 8.9, $T_s = 35.0$ °C, $p_s = 5.62$ kPa, shifted up 1.0 degC on the temperature axis and 0.32 kPa on the vapour pressure axis. These shifts equal the differences in skin temperature and saturated vapour pressure at skin temperature for the two subjects.

Standard isohids

Since a particular isohid can be adjusted for differences in skin temperature between different subjects, standard isohids constructed for one skin temperature can be used for all. Fig. 8.12 shows isohids for a standard skin temperature of 35 °C and heat

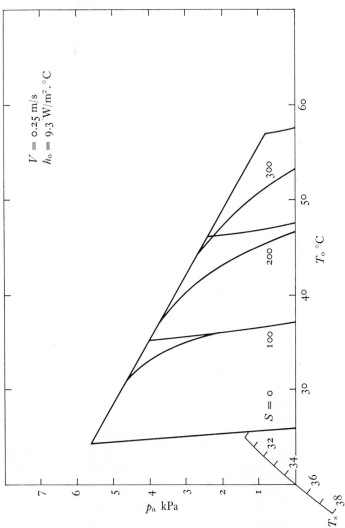

Fig. 8.12a. Standard isohids for skin temperature 35 °C, heat production 100 W/m². The isohid for some other skin temperature, T_s, can be obtained by moving the curve $(T_s - 35)$ °C to the right and $(p_s - 5.62)$ kPa upwards. The curve running through the intersection of the axes allows the adjusted isohid to be traced (see text). If the heat production is not 100 W/m² the isohid should be shifted $(H_p - 100)/1.05$ h_o °C to the left.

$V = 0.25$ m/s
$h_o = 9.3$ W/m².°C

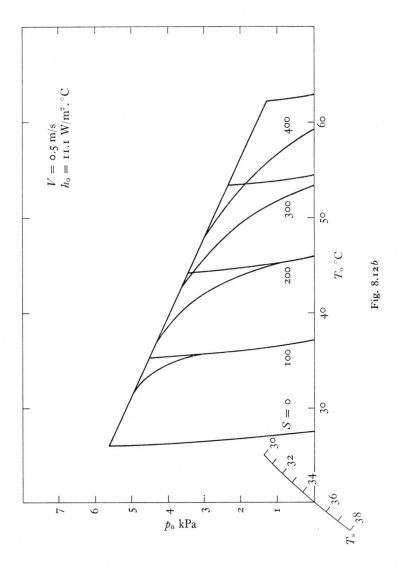

Fig. 8.12*b*

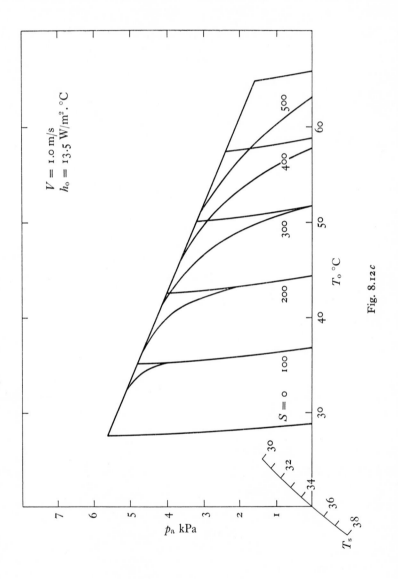

$V = 1.0\ \text{m/s}$
$h_0 = 13.5\ \text{W/m}^2 \cdot {}^\circ\text{C}$

p_a kPa

$S = 0$ 100   200   300   400   500

$T_0\ {}^\circ\text{C}$

T_s

Fig. 8.12c

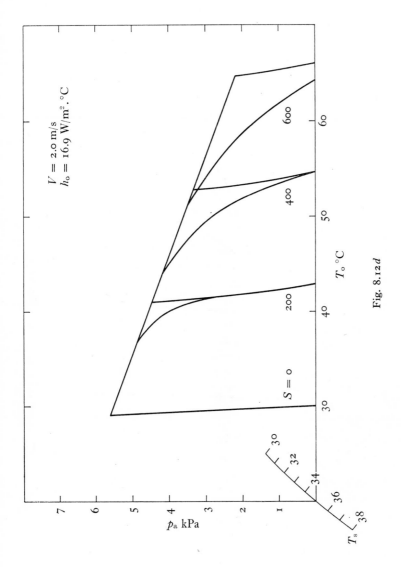

$V = 2.0$ m/s
$h_o = 16.9$ W/m^2 . °C

Fig. 8.12d

203

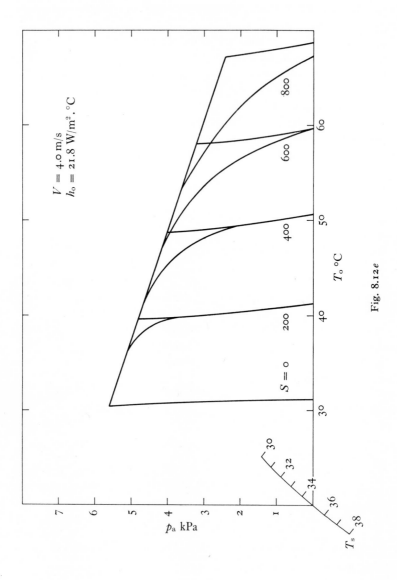

$V = 4.0 \text{ m/s}$
$h_o = 21.8 \text{ W/m}^2 \cdot {}^\circ\text{C}$

$S = 0$ 200   400   600   800

$T_o \, {}^\circ\text{C}$

p_a kPa

T_s

Fig. 8.12e

204

production of 100 W/m². Adjustment for the actual skin temperature is made by shifting the curve $(T_s - 35)$ degC to the right and $(p_s - 5.62)$ kPa upwards. The isohid for a chosen skin temperature (normal isohid) can be traced by using the short curved line running through the intersection of the axes. The axes are traced and their intersection is then located at the appropriate skin temperature on this line, keeping the axes parallel with those in the figure. The isohid can now be traced directly.

It is also possible to adjust for differences in the rate of heat production. If the respiratory minute volume were constant this could be effected by shifting the curve $(H_p - 100)/h_0$ degC to the left. In fact, when the rate of heat production increases so does the minute volume. The most obvious effect of this is on the line for full evaporation, but there is also a very small change in the slope of the wet skin line. The result is that the isohid is slightly altered in shape and is not shifted quite so far to the left as it would otherwise be. A shift of $(H_p - 100)/1.05h_0$ degC gives a good compromise at wind speeds from 0.25 to 4.0 m/s, sweat rates 0 to 400 W/m² and values of H_p from 50 to 200 W/m². The greatest errors (up to 0.5 degC in T_0) occur in very dry environments. If p_a is not less than 1.0 kPa the normal isohid shifted in this way fits the true isohid with an error of not more than 0.3 degC on the temperature axis. The accuracy with which environmental conditions, heat transfer coefficients and subject responses are likely to be known in practice is such that this error is insignificant.

Interpolation

Different isohids at the same wind speed are geometrically nearly similar in shape, though not in size. Any line drawn through the upper end of the wet skin line is intersected by the isohids at intervals nearly proportional to the sweat rates. Because the similarity in shape is not exact it is more accurate to interpolate between isohids 100 or 200 W/m² apart than to use a single master curve. Interpolation is illustrated in Fig. 8.13.

Operating conditions

The skin temperatures and sweat rates at which different subjects will reach heat balance under given conditions can be determined from the standard isohids and the subjects' skin temperature/sweat rate characteristics. The method is illustrated for the case

8. Equivalent environments – steady state

$H_p = 200 \text{ W/m}^2$, $T_0 = 38.0 \,°\text{C}$, $p_a = 3.00 \text{ kPa}$, $V = 1.0 \text{ m/s}$. It will be assumed that the efficiency of sweating varies with wettedness as it does for a uniformly sweating cylinder.

Standard isohids for the cylinder are shown in Fig. 8.13. The temperature and humidity of the environment are indicated by point A. The standard isohids are appropriate for a heat production

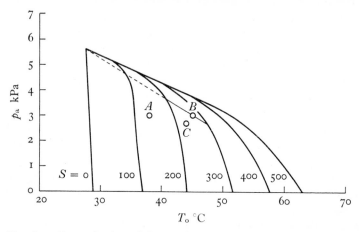

Fig. 8.13. Determination of the required sweat rate in the environment A, $T_0 = 38.0 \,°\text{C}$, $p_a = 3.0 \text{ kPa}$, $V = 1.0 \text{ m/s}$. The curved lines are standard isohids for a uniformly sweating cylinder, $T_s = 35.0 \,°\text{C}$, $H_p = 100 \text{ W/m}^2$. For a heat production of 200 W/m² the point A is shifted 7.05 deg C to the right (see text) to point B. The required sweat rate at $T_s = 35.0 \,°\text{C}$ is found by interpolation between the isohids for $S = 200$ and $S = 300$ along the line joining point B to the upper end of the wet skin line. Point C is for a skin temperature of 36.0 °C, and is 1.0 degC and 0.32 kPa lower than point B. A curve relating T_s to S_{req} can be produced in this way and is shown in Fig. 8.14.

of 100 W/m². For other rates they should be shifted to the left by $(H_p - 100)/1.05h_0$ degC, in this case 7.05 degC. It is simpler to shift point A to the right by this amount, as indicated by point B. The standard isohids relate to a skin temperature of 35 °C, and by interpolation the sweat rate required for heat balance at this skin temperature is 263 W/m². For a skin temperature of 36 °C the isohids are shifted up 1.0 degC and 0.32 kPa. It is simpler to shift point B down by these amounts, as indicated by point C. The appropriate sweat rate is 225 W/m². In this way a curve can be constructed which relates required sweat rate to skin temperature (Fig. 8.14). A subject will operate at the point on this curve which is consistent with his own skin temperature/sweat rate characteristic. For the subject illustrated the operating conditions are $T_s = 34.7\ ^\circ C$, $S = 270\ W/m^2$.

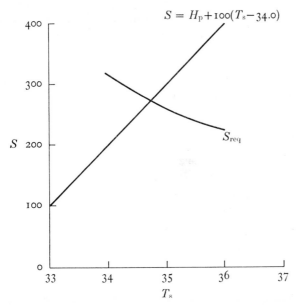

Fig. 8.14. Operating conditions for a subject in the environment shown in Fig. 8.13. The line for S_{req} is derived from that figure by the method described in the text. The sweating curve for a subject who sweats according to the equation $S = H_p + 100(T_s - 34.0)$ is shown by the straight line. The operating point for this subject is at the intersection of the two lines.

Hidromeiosis

Both the required sweat rate, derived from the normal isohids, and the subject's sweating curve are based on the assumption that there is no hidromeiosis. In practice if some parts of the skin are wet with sweat the sweat rate there will decline and the wastage of sweat will diminish. Hidromeiosis will stop when the sweat rate is equal to or a little less than the local maximum evaporative capacity. It will therefore have little effect on the rate of evaporation, and the skin temperature at which the subject operates will not be altered, nor will the central sweating drive. The normal isohids remain indicative of the heat stress, but if evaporation is limited the sweat rate will be less than that indicated by the isohid.

8. *Equivalent environments – steady state*

At any point on the skin surface the local maximum evaporative capacity is $\bar{E}_{\max} . h_e / \bar{h}_e$, where $\bar{E}_{\max}$ is the mean maximum evaporative capacity for the whole subject and $\bar{h}_e$ the mean evaporation coefficient. The distribution of the ratio $h_e / \bar{h}_e$ is known for the case of a cylinder in a transverse wind (Fig. 2.7), and the effect of hidromeiosis during a long exposure to heat can be calculated.

At any point on the surface if S_0, the sweat rate in the absence of hidromeiosis, exceeds the local maximum evaporative capacity the surface will be wet and the sweat rate will decline according to the equation

$$S_t = S_0 \, e^{-t/k}.$$

It will be assumed that this decline ceases when the sweat rate equals the local maximum evaporative capacity. The time, t_x, taken to reach this level is given by

$$t_x = -k . \ln \left(\frac{\bar{E}_{\max}}{S_0} . \frac{h_e}{\bar{h}_e} \right) .$$

If the total duration of the exposure, t_t, exceeds t_x, the total sweat produced from this part of the cylinder during the declining phase, S_d, is given by

$$S_d = k S_0 (1 - e^{-t_x/k}).$$

The total sweat produced during the final phase in which the sweat rate is equal to the local $E_{\max}$ is given by

$$S_f = (t_t - t_x) \, \bar{E}_{\max} \frac{h_e}{\bar{h}_e} .$$

If t_x is greater than t_t there is no final phase, and

$$S_d = k S_0 (1 - e^{-t_t/k}).$$

These equations show that the ratios S_d/S_0 and S_f/S_0 are functions of $(\bar{E}_{\max}/S_0)$, $(h_e/\bar{h}_e)$ and k. The total sweat loss is $(S_d + S_f)$ and for any point on the cylinder $(S_d + S_f)/S_0$ is a function of $\bar{E}_{\max}/S_0$. The total sweat loss from the whole cylinder can be found by integrating round the circumference.

For the purpose of predicting the total sweat loss during an exposure to heat some allowance must be made for the initial warming up period during which the sweat rate increases towards the value, S_0, required for heat balance in the absence of hidromeiosis. If we assume that the sweat rate increases linearly to reach this value at the end of the first half hour, and neglect hidromeiosis during the first hour, the total sweat rate during the first hour will be $0.75 \, S_0$. Fig. 8.15 shows the relation between the total sweat rate over four hours, S_4 (expressed as S_4/S_0), and $\bar{E}_{\max}/S_0$. The hidromeiosis time constant, k, is taken as 2 h.

If $\bar{E}_{\max}/S_0$ is greater than 2.20, no part of the surface is fully wet, there is no hidromeiosis and $S_4/S_0 = 3.75$. If $E_{\max}/S_0$ is less than 0.14 the sweat rate exceeds $E_{\max}$ at all points on the surface throughout the four hours and $S_4/S_0 = 2.30$.

From Fig. 8.15 and the normal isohids it is possible to construct lines of equal four-hour sweat rate. An example is shown in Fig. 8.16, for a subject with a heat production of 100 W/m² who sweats according to the equation,

$$S = H_p + 100(T_s - 34.0).$$

His normal isohids (standard isohids adjusted for skin temperature) are shown by the dashed lines. The continuous line shows environments in which, at a heat

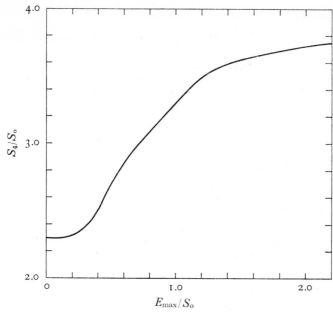

Fig. 8.15. The effect of hidromeiosis on the total sweat loss during a 4 h heat exposure. The assumptions upon which this curve is based are described in the text. S_4 is the total sweat loss during the 4 h expressed in $W.h/m^2$. S_0 is the sweat rate required for heat balance in the absence of hidromeiosis, i.e. at the subject's normal operating point for the environment and work rate in question.

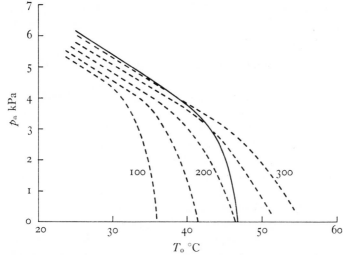

Fig. 8.16. Conditions under which a subject would sweat a total of 2 kg during a 4 h exposure to heat (continuous line). Heat production 100 W/m^2, wind speed 0.25 m/s. The subject sweats according to the equation $S = H_p + 100$ $(T_s - 34.0)$. Dashed lines show the appropriate isohids in the absence of hidromeiosis. The method of calculating the 4 h sweat loss is described in the text.

8. Equivalent environments – steady state

production of 100 W/m², he would sweat 2.0 kg during a 4 h exposure. This is equivalent to 740 W.h/m².

The point at which the four-hour line cuts the isohid for 300 W/m² is constructed as follows. $S_4/S_0 = 740/300 = 2.47$. From Fig. 8.15, E_{max}/S_0 is 0.37, so E_{max} is 111 W/m².kPa. The skin temperature is 36.0 °C, $p_s = 5.94$ kPa. At the prevailing wind speed h_e is 62 W/m².kPa. Hence p_a is 4.15 kPa.

Experimental observations

The above method of predicting the total sweat loss of a nude subject over 4 h is based on the following assumptions.

1. Heat exchanges by convection, evaporation and radiation proceed as described in Chapters 2 and 3. The various coefficients have been correctly measured.

2. The subject behaves like a uniformly sweating cylinder. It is necessary to assume this in order to determine the hidromeiosis effect, so this is the only case for which the four-hour sweat rate can be calculated.

3. In the absence of hidromeiosis the subject sweats according to the equation, $S = H_p + 100(T_s - 34.0)$. This is probably representative of moderately acclimatized young men (Hatch, 1963).

4. Hidromeiosis proceeds with a time constant of 2 h until the local sweat rate equals the local maximum evaporative capacity. Thereafter there is no further decline in sweating.

5. During the first hour the subject sweats 75 per cent of the rate required for heat balance in the absence of hidromeiosis.

On these assumptions the total sweat rate over 4 h can be calculated for any environment and work rate. The calculated values are compared in Fig. 8.17 with the observations of Dunham et al. (1946) (Macpherson, 1960, Tables 11 A and 13 A). Men wearing shorts only were exposed to environments in the range $T_0 = 32$–49 °C, $p_a = 2.4$–4.5 kPa, $V = 0.05$–3.35 m/s. They either rested or worked at a mean rate of 129 W/m². At very low air movements (below 0.25 m/s) the effect of movement of the subjects on the heat exchange coefficients is likely to be important and cannot be allowed for. Comparison of the observations with predicted values is therefore restricted to the wind speed range 0.25–3.35 m/s.

Except in the most severe environments at the higher work rate (the highest four observed sweat rates) agreement is reasonably good. The predicted value is about 7 per cent low, but the predic-

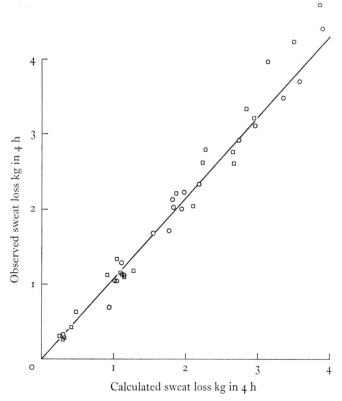

Fig. 8.17. Comparison of the 4 h sweat loss calculated by the method described in the text with the observations of Dunham *et al.* (1946).

tion is for nude subjects and so might be expected to be a little low in most circumstances. In the most severe environments the subjects were operating on or near the wet skin line. Here sweat rate has little or no effect on heat balance or skin temperature, provided, of course, that it is sufficient to wet the skin. With no physical feedback between sweat rate and skin temperature one would expect the sweat rates of different subjects to vary widely. Conversely, when there is full evaporation the sweat rate is not greatly influenced by the subjects' characteristics. An increase of skin temperature of 1 degC increases the sensible heat loss by h_0 W/m², reducing the required sweat rate by this amount. At a wind speed of 1.0 m/s this amounts to 0.14 kg in 4 h. Most of the examples fell

8. *Equivalent environments – steady state*

between these extremes and so constitute a good test of the propositions developed in this chapter.

Fig. 8.18 shows a similar comparison with the results of Adam *et al.* (1955) (Macpherson, 1960, Table 38). The environments and work rates were similar to those of Dunham *et al.* (1946), shown in Fig. 8.17. In this case the agreement is somewhat better.

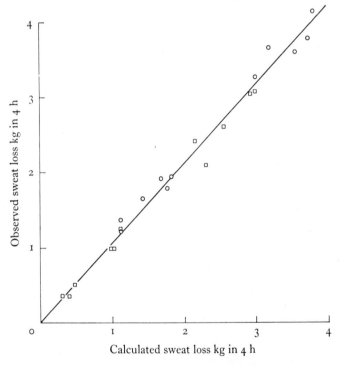

Fig. 8.18. Comparison of the 4 h sweat loss calculated by the method described in the text with the observations of Adam *et al.* (1955).

Calculation of the 4 h sweat loss was based on the standard isohids of Fig. 8.12. These were first adjusted for the assumed response characteristics of the subject (skin temperature/sweat rate relation), and a further adjustment then made for hidromeiosis. The agreement between the result of all these processes and the experimental observations is good (the residual variance is less than if the P4SR is used for the prediction), suggesting that the

principles of the approach are sound and that the isohids themselves are quantitatively correct. Without adjustment for hidromeiosis an isohid for a particular subject is also a skin isotherm
and therefore a line of constant stress and strain.

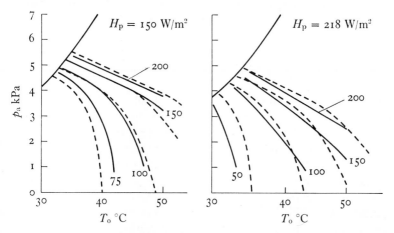

Fig. 8.19. Lines of equal E_p (index of physiological effect, Robinson *et al.* 1945).
Wind speed 0.92 m/s, heat production 150 W/m² (left-hand diagram) and
218 W/m² (right-hand diagram). The dashed lines are normal isohids for a subject who sweats according to the equation $S = H_p + 100(T_s - 34.0)$.

Robinson, Turrell & Gerking (1945) assessed the strain in subjects working or resting in the heat by adding functions of the
heart rate, skin temperature, rectal temperature and sweat rate.
These functions were based on measurements made at the appropriate work rate in a cool environment and in the most severe
environment tolerable for two hours. The total was called the index
of physiological effect, E_p. One may dispute the form of this index,
but since the physiological measurements are strongly correlated
with one another it is probable that conditions of equal E_p can be
regarded as imposing an equal strain.

Lines of equal E_p are shown in Fig. 8.19. The subjects wore
shorts only, and the wind speed was 0.92 m/s. The dashed lines are
normal isohids for a standard subject ($S = H_p + 100(T_s - 34.0)$) at
the appropriate work rate and wind speed. They correspond fairly
well with the E_p lines.

8. *Equivalent environments – steady state*

Formulation of strain

It appears reasonable to propose that at a given rate of heat production different subjects having the same skin temperature suffer an equal strain. Subjects may, however, differ in the way they deal with this strain. One may have a low core temperature and high body conductance, another the reverse. If the rate of heat production increases at given skin temperature, both core temperature and body conductance increase, probably differently in different subjects.

Hatch (1963) considers that the most important aspect of heat strain is the requirement for cutaneous blood flow, and suggests using the ratio of cutaneous blood flow to cardiac output as a single index of strain. The blood exchanges its heat a small distance below the skin surface, and Hatch bases his calculation of skin blood flow on the difference between core temperature and deep skin temperature, the latter being (H_s/K_s) degC above skin temperature, where K_s is the (constant) thermal conductance of the superficial layers of skin. This is a logical approach, but the strain cannot be stated without knowledge of the core temperature and cardiac output. If these are assumed, differences between different subjects are neglected and the index of strain becomes a mere compound of skin temperature and heat production. Hatch's strain index does not vary much with changes in heat production, because cardiac output is nearly proportional to heat production. Indeed if this were exactly so and the whole body conductance were used instead of the skin blood flow, the strain would be independent of heat production and proportional to $1/(T_c - T_s)$.

The observations of Saltin & Hermansen (1966) make it possible to compare thermal strain at different work rates in the prescriptive zone without the necessity of knowing the core temperature. The core temperature in the prescriptive zone (i.e. the core temperature at which subjects appear to prefer to operate) may be described by the equation

$$T_c = 36.5 + 3.0 \dot{V}_{O_2}/\dot{V}_{O_2}\text{max}.$$

It was pointed out in Chapter 7 that this type of relation would result if an imaginary layer between the core and the surface were maintained at 36.5 °C. The conductance deep to this layer is constant for any given subject. The conductance superficial to this layer is $H_s/(36.5 - T_s)$. In the prescriptive zone this can serve as an

214

indication of superficial blood flow. The resistance of the upper skin layers can be regarded as constant, as Hatch (1963) suggests. The effect of this is to place an upper limit on the possible value of the total conductance between the imaginary layer and the skin surface, but the value of this conductance remains an index though not a statement, of skin blood flow.

This index of strain is the superficial conductance required to maintain the core temperature at its preferred level, whatever that level may be. Within the prescriptive zone it increases with skin temperature and with heat production, becoming infinite when the skin temperature is 36.5 °C. It has the advantage of providing a plausible way of combining skin temperature and heat production without introducing characteristics which may differ between different subjects. It is inadequate in that it does not cope with situations in which the skin temperature is greater than 36.5 °C. To maintain the imaginary layer at 36.5 °C would then require a negative conductance, the magnitude of which diminishes with increasing skin temperature or heat production. The same negative conductance would be required at $T_8 = 37.0$ °C, $H_8 = 50$ W/m^2 as at $T_8 = 37.5$ °C, $H_8 = 100$ W/m^2, and it is clearly absurd to suppose that the strain is the same in these cases. The tolerance of subjects to conditions beyond the prescriptive zone is very variable, and it is probable that physiological properties such as maximum skin blood flow, which have not so far entered this discussion, become important. It may therefore be wise to describe strain merely by stating skin temperature and rate of heat production, without attempting to combine these factors.

Clothing

No mention has so far been made of the effects of clothing. The heat exchange coefficients used in the development of the isohids have been those appropriate for nude subjects, and the experimental observations quoted have been made on subjects wearing, at the most, shorts and shoes. In industrial and military situations it is commonly necessary for clothing to be worn, and an index of heat stress which cannot be applied to the clothed subject is almost valueless.

The survey of the thermal properties of clothing in Chapter 5 indicates both the complexity of the problem and the dearth of

8. *Equivalent environments – steady state*

quantitative information about it. The effect of clothing on sensible heat exchange is in principle fairly simple, and could be expressed by a net heat exchange coefficient and an appropriately modified Operative temperature. However, both these quantities depend on the ambient air movement and the postural activity of the subject as well as on the nature of the clothing assembly. In the present state of knowledge they cannot be deduced from the properties of the garments, but must be measured empirically. Moreover, if sweat wets the clothing, as it does in situations where there is an important heat stress, the exchange of sensible heat through the clothing is modified. The effects of clothing on evaporation are more complicated, principally because of the wicking of sweat onto the fabric, and also because the effective evaporation coefficient for ventilating air depends on the sweat rate. The methods used for assessing heat stress for the nude subject cannot be applied if the heat exchange coefficients depend on the sweat rate.

The present state of knowledge is such that it appears impossible to include all the effects of clothing in a rationally based index of heat stress. The information required about any one assembly would involve observations over such a wide range of environments and work rates that the results would provide an adequate empirical statement by themselves.

Humidity and sensation

In the above discussion of equivalent environments the only relations between subject and environment considered have been those of heat transfer. The stress of the environment was considered to be the impediment to heat loss by the subject, and the strain the internal adjustments the subject has to make in order to maintain heat balance. This led to the conclusion that if the wettedness is less than about 0.5 the effect of humidity is small, being due only to differences in the heat of evaporation of sweat, insensible perspiration and evaporation from the respiratory tract.

Common experience suggests that the effect of humidity is subjectively greater than this and that environments which make equal demands on the circulation and sweating mechanism may not necessarily be judged equally unpleasant. Wetting of the skin and clothing with sweat is certainly a source of discomfort (Gagge,

Stolwijk & Saltin, 1969), but the 'oppressiveness' of a humid environment is apparent under conditions in which there is no obvious limitation of evaporation. The subjective sensation of a warm humid environment cannot be reproduced in a drier, hotter environment, and the latter may well be judged pleasanter even if skin temperature and sweat rate are higher. The moderate heat stress of sunbathing appears to be a positive pleasure, unlike the admittedly more severe stress of the Turkish bath, endured for its after-sensation.

It appears, therefore, that 'mugginess' may be sensed in some way, and that it may contribute to the unpleasantness of warm environments. If so, the subjective impression of an environment has at least two components, one depending on the requirements for heat balance, the other on direct sensations of humidity. Since these differ in character it may not be possible to compare them directly. Although the sensations in two situations cannot be equated, a subject may prefer one to another, and equivalence of the two can be said to exist when he finds it impossible to choose between them.

The relation between subjective comfort and ambient temperature, humidity and air movement has been the subject of many investigations (Houghten & Yagloglou, 1923; 1924; Yaglou & Miller, 1925; Winslow, Herrington & Gagge, 1937c; Rowley, Jordan & Snyder, 1947; Missenard, 1948; Koch, Jennings & Humphreys, 1960). In warm environments humidity is reported to have an adverse effect on comfort. This is evident under conditions in which no loss of sweat by dripping either is reported or would be expected. At given temperature and humidity subjects often prefer a high air movement, even when air temperature is above skin temperature.

A sensation of humidity could be derived from either the skin or the respiratory tract. In the latter case it would be independent of air movement. When air temperature exceeds skin temperature and evaporation is complete, increasing air movement must increase the heat gained from the environment, and skin temperature and sweat rate must be higher. A sensation of humidity which is independent of air movement cannot account for the subjective preference for higher air movement, so the respiratory tract does not appear to be the primary source of this sensation.

Winslow, Herrington & Gagge (1937c) found that comfort was

better correlated with wettedness than with either skin temperature of body conductance. Wettedness depends on sweat rate, ambient humidity and air movement, and at a fixed rate of heat production in the steady state, sweat rate depends on the heat gain from the environment. Thus wettedness combines ambient humidity and sensible heat stress, and, if both these contribute to the subjective assessment, might well be the best single correlate of those which were tried.

It is possible that the sensation of humidity depends on skin relative humidity. The swelling of epidermis is a function of its relative humidity (Fig. 7.12) as is that of hair (Gregory & Rourke, 1957). Alteration in the mechanical properties of the epidermis and hairs might affect cutaneous nerves, particularly those around the hair follicles, and tactile sensations are known to affect impressions of freshness (Bedford, 1964, p. 132).

In making a preference judgement between two environments a subject may trade skin temperature (i.e. sensible heat stress) against skin relative humidity. Changing the air movement will affect both these factors, but to different extents and possibly in different directions. Consider two environments of equal temperature ($T_a = \bar{T}_r = T_0$) and vapour pressure. The air movement in one, V_1, is lower than that in the other, V_2. In neither case is sweat lost by dripping. If the air temperature is above skin temperature, increasing the air movement must increase the sensible heat stress, so the skin temperature, skin blood flow and sweat rate must be higher at V_2 than at V_1. If the subject compares the environments only on the basis of these factors, he will judge V_2 to be more unpleasant than V_1. The skin relative humidity is given by

$$\phi_s = \frac{p_a}{p_s} + \frac{H_s}{h_e . p_s} + \frac{h_0}{h_e . p_s}(T_0 - T_s). \tag{5}$$

At the higher air movement, h_e will be larger. Since the sensible heat gain is greater, both T_s and p_s will be greater. The first two terms on the right-hand side of the equation must therefore be smaller. In the last term $(T_0 - T_s)$ will be smaller, as will the ratio h_0/h_e, since although both coefficients increase with increasing air movement h_0 contains the constant term, h_r, and its proportionate increase is less. Thus all the terms become smaller at higher air movement, and ϕ_s must be smaller at V_2 than at V_1. If the subject compares the environments only on the basis of skin relative humidity he will judge V_1 to be more unpleasant than V_2. If his judgement is based on some combination of sensible heat stress and skin relative humidity the subjective effect of changes in air movement will depend on the ambient temperature and humidity. In hot dry environments a lower air movement may be preferred, whereas in warm humid conditions the predominant effect of air movement is on ϕ_s, and a higher air movement may be preferred. These general preferences accord with common experience, and are given quantitative expression in the Effective temperature and Equivalences en séjour nomograms. These are considered in more detail in the next chapter, where the latter is shown to be consistent with the proposition that subjective preference is based on both skin temperature and skin relative humidity.

Another possibility is that even when the mean wettedness is low, evaporation may be limited by ambient conditions in some regions where the local sweat rate is high, for example, the forehead (Ferguson *et al.* 1956). The fact that sweat does not run off cannot be taken as evidence that evaporation is not limited by ambient conditions, since hidromeiosis might adjust the local sweat rate to match the rate of evaporation. Assuming that this occurs on some parts of the face, the skin temperature there will depend on the ambient humidity. Furthermore the total area of skin from which evaporation is limited will increase with increasing humidity. Increasing the air movement will lower the temperature and decrease the extent of the region from which evaporation is limited. If subjective assessment of the environment is predominantly influenced by the condition of the face, this would explain the preference for low humidity and high air movement in situations where no sweat is lost by dripping.

An increase of forehead temperature with ambient humidity was observed by Sheard, Williams & Horton (1939) at ambient temperatures as low as 25.0 °C. Mean results for two subjects are shown in Fig. 8.20. It is noteworthy that the mean temperature of the fingers and toes decreased with increasing humidity. This may be because diminished evaporation from the face was compensated for by increased sweating from regions where evaporation was not limited by environmental factors.

It is probable that the sensation from regions of limited evaporation will be related to the mean wettedness of the whole subject, W. Clearly the extent of these regions will increase with W (p. 45).

8. Equivalent environments – steady state

Their temperature can be calculated if some simplifying assumptions are made. The heat balance equation for the wet forehead is

$$k_f(T_c - T_f) = h_{of}(T_f - T_0) + h_{ef}(p_f - p_a), \qquad (6)$$

where the subscript 'f' denotes conditions for the forehead. T_c is

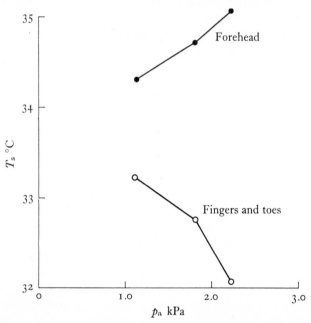

Fig. 8.20. Influence of humidity on the skin temperatures of the forehead, ●, and the fingers and toes, ○. Mean results for two subjects at ambient temperatures of 25.0, 26.1 and 27.2 °C. (Data from Sheard *et al.* 1939.)

the body core temperature and k_f the conductance of the tissues of the forehead. Let it now be assumed that

$$h_{of}(T_f - T_0) = x\bar{h}_0(\bar{T}_s - T_0),$$
$$h_{ef}(p_f - p_a) = y\bar{h}_e(\bar{p}_s - p_a),$$

where x and y are constants and the superscript bar denotes mean values for the whole body. Neither equation is exactly true, but the error may not be too great if T_f is fairly close to $\bar{T}_s$, since h_{ef} is proportional to $\bar{h}_e$, and h_{of} roughly proportional to $\bar{h}_0$.

220

Now $\bar{h}_o(\bar{T}_s - T_o) = S - H_p$, and $\bar{h}_e(\bar{p}_s - p_a) = E_{\max}$, so equation (6) can be expressed

$$k_f(T_c - T_f) = x(S - H_p) + yE_{\max},$$

$$k_f T_f = k_f T_c + xH_p - xS\left(1 + \frac{y}{xW}\right).$$

It is not unreasonable to assume that k_f is constant (Froese & Burton, 1957). At constant work rate in the prescriptive zone both H_p and T_c are constant and S is a function of $\bar{T}_s$. Hence T_f is a function of $\bar{T}_s$ and W and this might account for the high correlation between comfort and wettedness reported by Winslow, Herrington & Gagge (1937c).

9 INDICES OF HEAT STRESS – STEADY STATE

There have been many attempts to devise ways of reducing the thermal components of an environment to a single figure indicative of the heat stress or of the likely condition of an individual when he has reached the steady state. The possible physical bases for such systems have been examined in the preceding chapters, and it is evident that a family of environments which are equivalent (on whatever criteria) for one subject at one work rate will not necessarily be equivalent for another subject or a different work rate. No index in which environmental factors are combined into a single number can be appropriate for all subjects and work rates, and there is no simple way whereby the physiological characteristics of the subject may be combined with the physical factors of the environment to yield a single figure. A solution is possible in principle (Chapter 8), but is too cumbersome for practical application.

There is a need in industry and in the armed services for some general indication of the comparative stresses of different warm environments. Such an index of heat stress must be simple enough for everyday use, and something must be sacrificed in order to achieve this. If the index covers a wide range of conditions, differences between subjects will be important, and it is necessary either to accept a substantial loss of accuracy or to restrict the index so that it applies only to certain categories of subjects. In some industrial situations the range of humidity at any given temperature may be quite small, and when the envelope of conditions is restricted in this way the rank order of stress of different environments is less dependent on the responses of the subject. If air movement is constant and humidity and temperature increase together, as for instance if the air is always saturated, subject responses have no effect on the rank order; the higher the temperature the greater the stress. Here there is no need to combine temperature and humidity and the problem of constructing the index vanishes. Differences in individual heat tolerance can readily be examined in such a restricted range of conditions, and recommendations of work routine and personnel selection can be made with confidence (Wyndham *et al.* 1967*b*).

Some of the more important indices of heat stress are reviewed below and related to the analysis presented in the earlier part of this book. Others are listed by Fanger (1970). The indices considered here fall into three groups:

1. Indices based on analysis of heat exchange:
   a. Heat stress index of Belding & Hatch (HSI),
   b. Index of thermal stress (Givoni) (ITS).
2. Indices based on physiological observations:
   a. Predicted four-hour sweat rate (McArdle) (P4SR),
   b. Wet bulb globe temperature index (Yaglou) (WBGT).
3. Indices based on subjective preference:
   a. Effective temperature (Yaglou) (ET),
   b. Equivalences en séjour (Missenard) (ES).

The heat stress index of Belding & Hatch (HSI)

This index (Belding & Hatch, 1955) is based on the physical analysis of heat exchange developed by Machle & Hatch (1947) and later applied to industrial situations (Haines & Hatch, 1952). Essentially two quantities are estimated, the rate of evaporative heat loss required for heat balance, E_{req}, and the maximum evaporative capacity, E_{max}. The number describing the heat stress is the ratio E_{req}/E_{max}, expressed as a percentage. In the steady state the rate of evaporation equals E_{req}, and the ratio E_{req}/E_{max} is the wettedness. (An arbitrary upper limit of 380 W/m² is placed on E_{max}, and this value is used for the calculation if the actual value is greater.)

In order to simplify the index and to make it independent of the responses of the subject, all calculations are based on a skin temperature of 35.0 °C and a body surface area of 1.86 m². Respiratory heat exchange is neglected. These restrictions allow E_{req}, E_{max} and their ratio to be found from simple nomograms of the form of Fig. 9.1. In the original nomogram a psychrometric chart is included so that p_a can be found from wet and dry bulb observations. The heat exchange coefficients used in the original nomogram were based on the work of Nelson *et al.* (1947) on nude subjects.

9. Indices of heat stress – steady state

Hatch (1963) has since proposed modified formulae in which both h_c and h_e are proportional to $V^{0.6}$.

When $E_{req} = E_{max}$ the HSI is 100. At larger values the rate of evaporation required for heat balance (at a skin temperature of 35.0 °C) exceeds the maximum evaporative capacity. Either the subject is not in heat balance or his skin temperature must be substantially higher than 35.0 °C.

It is perhaps remarkable that an index of heat stress intended for industrial application should be based on measurements made on nude subjects, but the authors considered that the errors introduced by a single layer of light clothing were unlikely to be serious. In oral discussion Belding stated that the calculated values of stress coincided quite well with observed strains in men wearing light clothing, and suggested that this agreement was a consequence of assuming a skin temperature of 35 °C, a degree or more lower than the likely values under high stress conditions (Belding & Hatch, 1956). More recently, Hertig & Belding (1963) have proposed that for subjects wearing normal work clothing the values of h_o and h_e should be reduced by 30 per cent. The nomogram shown in Fig. 9.1 uses Hatch's (1963) proposals for the nude case, and these have been reduced by 30 per cent for the clothed case. The original nomogram is available from several sources (Belding & Hatch, 1955; 1956; Lind, 1964, p. 69 (°F, ft/min, B/h); Löfstedt, 1966 (°C, m/s, kcal/h)).

Subject to the simplifying assumptions listed earlier, the nomogram allows both E_{req} and E_{req}/E_{max} to be estimated. The latter is taken as the measure of heat stress (HSI) and for index values below 100 is identical with the wettedness at $T_s = 35.0$ °C. It has been objected that it is unreasonable to equate the stresses on two subjects merely on the ground that their wettedness is the same. Despite a low sweat rate the wettedness may be large, and it is unrealistic to describe the stress on a subject in this state as severe (Hick, 1956). It would seem more reasonable to combine E_{req} and W in some way which would allow both to enter the resulting expression of stress. In Chapter 8 they are combined in the calculation of S_{req} for the nude subject, and Givoni (1963) has used the same approach for the clothed subject. Givoni's index (see below) is based on an assumed skin temperature of 35 °C and can be regarded as a logical extension of the HSI.

At high stresses the wettedness predicted by the HSI exceeds

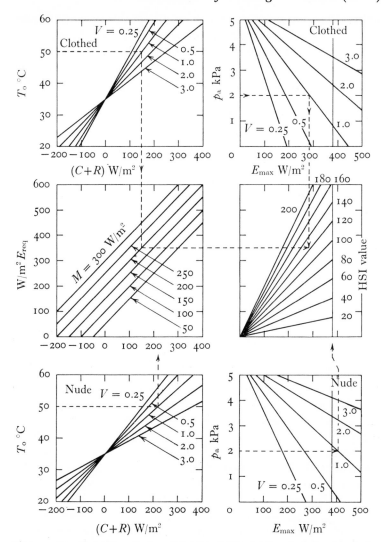

Fig. 9.1. The Heat Stress Index of Belding & Hatch. For clothed subjects the upper two diagrams are entered with Operative temperature, T_o, and ambient water vapour pressure, p_a, as shown. The HSI value is given in the right-hand middle diagram. For nude subjects the nomogram is entered through the bottom two diagrams as indicated. If E_{max} exceeds 380 W/m² the final diagram is entered with this figure, as shown for the nude case, $p_a = 2.0$ kPa, $V = 1.0$ m/s. The heat exchange coefficients used in constructing this nomogram are those recommended by Hatch (1963), with the adjustment for clothing proposed by Hertig & Belding (1963).

9. *Indices of heat stress – steady state*

unity. At the assumed skin temperature the subject could not be in heat balance. The actual rate of evaporation would be $E_{\max}$ and the subject would store heat at a rate $(E_{\mathrm{req}} - E_{\max})$. This is equal to $E_{\max}(HSI - 100)/100$, so environments of equal index value do not necessarily imply equal rates of heat storage.

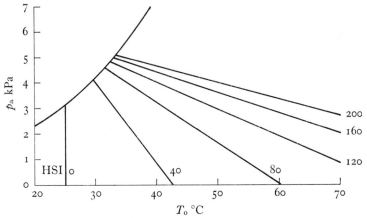

Fig. 9.2. Environments of equal HSI plotted on the psychrometric axes. The subject is clothed: $H_\mathrm{p} = 100$ W/m², $V = 1.0$ m/s.

The Belding & Hatch index sidesteps the problem of subject responses by assuming a constant skin temperature. This allows it to be represented as a pure index of stress, depending only on environmental factors. There is a logical difficulty here, since in order to achieve this the condition of the subject (skin temperature) has been assumed to be independent of the stress. One cannot then say that strain is a function of stress. This objection can be dismissed as pedantic provided that the index arranges environments correctly in rank order. However, it must fail to do this, for some subjects at least, since it defines equivalent environments without reference to the subject's responses. Environments which are equivalent for one subject are not necessarily so for another.

At given air movement environments of equal HSI fall on straight lines on the psychrometric axes. This is shown in Fig. 9.2 for the clothed subject at an air movement of 1.0 m/s, heat production 100 W/m². For the nude subject the lines are displaced 3.0 degC on the temperature axis. Except for very rough purposes these lines cannot be regarded as providing a realistic statement of heat stress, since lines of equivalence based on physiological observations (E_{p}, P4SR) or on physical analyses (Woodcock, Pratt & Breckenridge, 1952; Givoni, 1963; and Chapter 8 of this book) are usually far from straight.

In view of the large variations between subjects, and because in practice their individual or group characteristics are often unknown, the accuracy required of a heat stress index for industrial application is frequently small. The Belding & Hatch index is perhaps adequate for many purposes, and has the merit of being based on physical principles. The components contributing to the heat stress become apparent when the index is applied, and this is helpful in devising improvements in the working environment.

Index of thermal stress (ITS)

In 1963 Givoni described an 'Index of Thermal Stress' which predicted the sweat rate required for thermal equilibrium in subjects wearing summer clothing (see also Givoni, 1964). This was later extended to cover other clothing assemblies and to include an allowance for solar radiation (Givoni & Berner-Nir, 1967a). The fundamental structure of the index is rational. As in the HSI, E_{req} and E_{max} are first estimated. The next step is to take their ratio and to find from this the efficiency of sweating, η_{sc}. The index value is the sweat rate required for thermal equilibrium, $E_{\mathrm{req}}/\eta_{\mathrm{sc}}$. This appears to be the most logical way of combining sensible heat stress and evaporative capacity, but its strict application, even to the simplest case of the nude subject, is a complicated affair, and it may not be correct to assume that η_{sc} can be defined solely by $E_{\mathrm{req}}/E_{\mathrm{max}}$ (Chapter 5).

Givoni has made the ITS simple and workable by an astute blending of theory with empiricism. The formulae for the calculation of E_{req} and E_{max} are based on an assumed skin temperature of 35.0 °C, which makes the arithmetic (and the algebra) easy. The coefficients are empirical, and since they are derived by fitting experimental observations to equations based on a skin temperature of 35 °C much of the error inherent in assuming a skin temperature disappears. Suppose that at a given value of k_0 (the coefficient for sensible heat exchange between skin and environment) the skin temperature varies with T_0 according to the equation

$$T_{\mathrm{s}} = 35 + b(T_0 - 35).$$

9. Indices of heat stress – steady state

The rate of sensible heat gain, H_0, is $k_0(T_0 - T_s)$. By substitution

$$H_0 = k_0(1-b)(T_0 - 35).$$

The empirically fitted coefficient would be equal to $k_0(1-b)$ and would thus allow for the subject's responses as well as for the clothing. In contrast, the HSI uses the true coefficient, k_0, and so leads to erroneous estimates of H_0.

For the sake of consistency Givoni's formulae are here converted into s.i. units, and rates of heat transfer are expressed per unit area, assuming a DuBois area of 1.8 m². The sensible heat gain is given by

$$H_0 = \alpha \, . \, V^{0.3}(T_g - 35), \tag{1}$$

T_g is the globe temperature, roughly equal to the Operative temperature. In sunlight T_a is substituted for T_g, since separate allowance is made for solar radiation. The coefficient, α, depends on the clothing. Values for three clothing assemblies are given in Table 9.1. The form of equation (1) bears little relation to the physical processes of convection and radiation, but the expression has to fit results from different clothing assemblies and to make due allowance for skin temperature. Laboratory determinations of heat exchange coefficients are rather variable, even for nude subjects (Fig. 2.3), and the simple form of equation (1) is probably adequate for application to field conditions.

The solar radiation load is expressed by the equation

$$R_s = I_N \, . \, K_{pe} \, . \, K_{cl}(1 - a(V^{0.2} - 0.88)), \tag{2}$$

I_N is the normal solar intensity, K_{pe} a coefficient depending on posture and terrain, and K_{cl} and a coefficient depending on the clothing. Like the ITS itself this expression is a blend of theory and empiricism. Its derivation is explained by Givoni & Berner-Nir (1967b) and the values of the coefficients are based on their own experimental work (Givoni & Berner-Nir, 1967c). The factors K_{cl} and a describe the effect of clothing on solar heat load. For the lightest clothing assembly one would expect K_{cl} to be about 1.0, as it is, but the value of $a = 0.35$ is unexpected. The differences in heat gain between sun and shade should be independent of air speed for the nude subject, so a should be zero. However solar radiation raises the skin temperature, decreasing H_0 in equation (1). It is not possible to allow for this in equation (1), but since H_0 and R_s are added in order to obtain the total sensible heat gain, an appropriate empirical allowance in equation (2) will serve instead.

Clothing acts as a radiation screen, and the heat gain from solar radiation is reduced in the ratio k_0/h_0 (p. 107). Equation (2) introduces two terms to describe this. K_{cl} is the ratio appropriate for a wind speed of 0.5 m/s, taken as a base because this is a reasonable minimum for outdoor conditions. (The number 0.88 in equation (2) is $0.5^{0.2}$.) With increasing wind speed both k_0 and h_0 increase, but their ratio changes less than one might perhaps expect. Assuming a constant clothing insulation of 0.1 °C.m²/W, the ratio k_0/h_0 is 0.47 at $V = 0.5$ m/s, and 0.31 at $V = 4.0$ m/s. Over this range of wind speed h_0 doubles. The rather small effect of wind speed is reflected in the low exponent (0.2) in equation (2). The value of $K_{cl}(1 - a(V^{0.2} - 0.88))$ for a subject wearing military overalls falls from 0.40 at $V = 0.5$ m/s to 0.31 at $V = 4.0$ m/s.

The required evaporative heat loss, E_{req}, is the sum of the heat production, H_p, and the heat gain from the environment, $H_0 + R_s$.

The equation for E_{max} contains one coefficient, p, which depends on clothing.

$$E_{max} = pV^{0.3}(5.62 - p_a). \tag{3}$$

The formula for η_{sc} is independent of clothing

$$\eta_{sc} = \exp\left[-0.6((E_{req}/E_{max}) - 0.12)\right]. \tag{4}$$

The required sweat rate (ITS value) is

$$(H_p + H_0 + R_s)/0.37\eta_{sc} \text{ g/h}.$$

The factor 0.37 converts W/m² to g/h for a subject of area 1.8 m². If E_{req}/E_{max} is less than 0.12, η_{sc} is taken as 1.0; if it exceeds 2.15, η_{sc} is taken as 0.29. Extension of this relation to values of E_{req}/E_{max} greater than 1.0 is a way of compensating for the assumption of a constant skin teperature in equation (3). This cannot be done by modifying p (as α was modified in equation (1)) because skin temperature is not directly affected by E_{max}. Values of E_{req}/E_{max} as high as 2.15 are not unreasonable. If p_a is 5.30 kPa and the actual skin temperature is 36 °C ($p_s = 5.94$ kPa), the rate of evaporation from wet skin would be $0.64h_e$. Calculated on the basis of a skin temperature of 35 °C ($p_s = 5.62$), E_{max} would be $0.32h_e$.

The absence of a clothing coefficient from equation (4) makes it possible in principle to determine α and p for any clothing assembly from only two experiments. The first is conducted under conditions of low humidity and low sweat rate, when η_{sc} can be assumed

9. *Indices of heat stress – steady state*

to be unity. Measurement of heat production and sweat rate at one wind speed and air temperature allows α to be found. A second experiment conducted at high humidity and high sweat rate provides a measure of S_{req} for these conditions. E_{req} can be calculated since α is now known, and this allows η_{sc} to be found. Because the relation between η_{sc} and E_{req}/E_{max} is independent of clothing, E_{max}, and hence p, can be established.

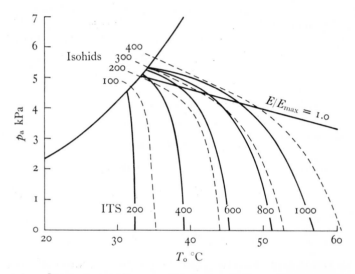

Fig. 9.3. Environments of equal ITS plotted on the psychrometric axes. The subject is seminude; $H_p = 100$ W/m², $V = 1.0$ m/s. The dashed lines are isohids for a nude subject who sweats according to the equation $S = H_p + 100$ $(T_s - 34.0)$. The straight line shows conditions in which the ratio E_{req}/E_{max}, calculated by the formulae used in the ITS, is unity.

The continuous lines in Fig. 9.3 are lines of equal ITS for the seminude subject (bathing suit and hat), $H_p = 100$ W/m², $V = 1.0$ m/s. The dashed lines are normal isohids (not corrected for hidromeiosis) derived by the methods of Chapter 8, for a nude subject who sweats according to the equation

$$S = H_p + 100(T_s - 34.0).$$

The two sets of curves are similar in general shape and distribution, showing that the ITS leads to much the same result as the more complicated methods of Chapter 8. Quantitative agreement between the two methods of predicting the required sweat rate is close. For a subject of surface area $1.8\ m^2$ a sweat rate of 800 g/h is equivalent to 296 W/m². The line for ITS 800 corresponds closely with the isohid for 300 W/m².

The ITS has been tested against the results of five experimental series (Dunham *et al.* 1946; Macpherson, 1960; Givoni & Rim, 1962; Kraning, Belding & Hertig, 1966; Givoni & Berner-Nir,

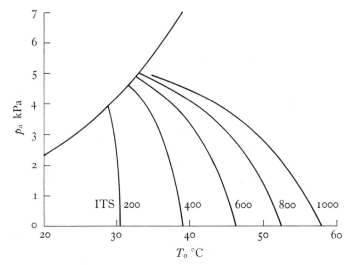

Fig. 9.4. Environments of equal ITS plotted on the psychrometric axes. The subject is wearing overalls; $H_p = 100$ W/m², $V = 1.0$ m/s.

1967 *a*). In general it gives a good quantitative prediction of required sweat rate, the observations being scattered about by ±20 per cent around the line of equality (Givoni & Berner-Nir, 1967 *a*). A very wide range of environments and clothing is covered, and this remarkably good agreement has been achieved without sacrificing the simple and workable structure of the index.

Fig. 9.4 shows lines of equal ITS for a subject wearing military overalls, $H_p = 100$ W/m², $V = 1.0$ m/s. These lines have the same general form as those of equal P4SR (Fig. 9.7), and seem more likely to be a true representation of conditions of equal heat stress than the straight lines of the HSI.

9. Indices of heat stress – steady state

TABLE 9.1. *Quantities used in calculating the ITS. The coefficients have been adjusted so that* H_o, R_s, E_{req} *and* E_{max} *are expressed in* W/m^2, *and vapour pressure in kPa. The final ITS value is in* g/h, *and the necessary conversion factor is included in the tabulated quantity* F_η.

V	$V^{0.3}$	$V^{0.2}$	E_{req}/E_{max}	$F_\eta (= 1/0.37\eta_{sc})$
0.2	0.62	—	0.2	2.8
0.4	0.76	—	0.3	3.0
0.6	0.86	0.90	0.4	3.2
0.8	0.94	0.96	0.5	3.4
1.0	1.00	1.00	0.6	3.6
1.2	1.06	1.04	0.7	3.8
1.4	1.11	1.07	0.8	4.1
1.6	1.15	1.10	0.9	4.3
1.8	1.19	1.13	1.0	4.6
2.0	1.23	1.15	1.1	4.9
2.5	1.32	1.20	1.2	5.2
3.0	1.39	1.25	1.3	5.5
3.5	1.46	1.28	1.4	5.8
4.0	1.52	1.32	1.5	6.2
4.5	1.57	1.35	1.6	6.6
5.0	1.62	1.38	1.7	7.0
5.5	1.68	1.41	1.8	7.4
6.0	1.71	1.43	1.9	7.9
6.5	1.75	1.45	2.0	8.4
7.0	1.79	1.48	2.1	8.9
			2.15	9.2

Clothing coefficients

Clothing	α	K_{cl}	a	p
Bathing suit and hat	10.2	1.0	0.35	153
Light summer clothing, underwear, short sleeved cotton shirt, long cotton trousers, hat	8.4	0.5	0.52	63
Military overalls over shorts	7.5	0.4	0.52	63

Solar coefficient for posture and terrain

Posture	Terrain	K_{pe}
Sitting, back to sun	Desert	0.39
	Forest	0.38
Standing, back to sun	Desert	0.31
	Forest	0.27

Equations

$$H_o = \alpha . V^{0.3}(T_g - 35)$$
$$R_s = I_N . K_{pe} . K_{cl}(1 - a(V^{0.2} - 0.88))$$
$$E_{max} = p . V^{0.3}(5.62 - p_a)$$
$$E_{req} = H_p + H_o + R_s$$
$$ITS = F_\eta . E_{req}$$

The predicted four-hour sweat rate (P4SR)

During the latter part of the second world war a series of studies on the physiological effects of hot climates was undertaken by the Medical Research Council in London on behalf of the Royal Navy. The four-hour duration of a naval watch was adopted as the standard experimental period and the subjects were naval ratings who had been acclimatized to moderate work in the heat. The P4SR index was developed from these experimental data (McArdle *et al.* 1947).

The index combines the effects of air temperature, radiant temperature, humidity, air movement and rate of working for two clothing assemblies, shorts and overalls (worn over shorts). The index number is the total sweat loss (litres) during a four-hour exposure. The nomogram (Fig. 9.5) was constructed by trial and error with the aim of providing the best possible indication of heat stress. Factors other than sweat rate were taken into consideration. For example at a given sweat rate the rectal temperature was found to be less elevated in hot dry than in humid warm conditions. The nomogram was arranged to underestimate the sweat rate in the former case thus giving a better indication of strain and therefore of stress. In severe conditions (P4SR greater than 5 l) sweat rate is deliberately overestimated. The actual sweat rate must be near its maximum and no longer reflects changes in heat stress adequately.

The use of sweat rate as an index of stress rather than strain has been questioned, and modifications have been proposed in order to remove this objection (Lind, 1964, p. 66). However, the index was developed from observations on a restricted class of subjects whose physiological responses are built into it, and the distinction between stress and strain need not be drawn. A more important point is that sweat rate may not be an adequate indication of either. The deliberate distortions referred to earlier were introduced in order to allow for this, and they compensate to some extent for the effects of hidromeiosis in humid environments and for saturation of the sweat mechanism in severe conditions.

The P4SR nomogram is shown in Fig. 9.5. Here and in the instructions for using it which follow, the units have been converted to °C, m/s, and W/m². Interpolation is difficult and may be unjustifiable in an empirical nomogram. The wind speeds used in the original nomogram have therefore been retained despite their

9. *Indices of heat stress – steady state*

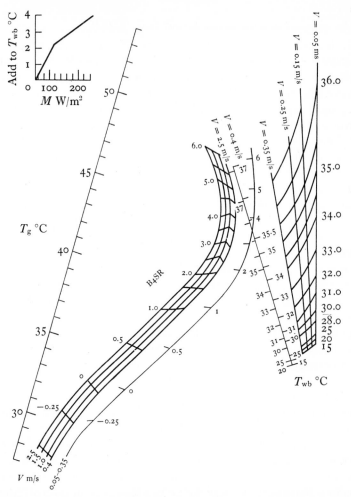

Fig. 9.5. The P4SR nomogram. The left-hand scale is entered with the globe temperature, and the appropriate wet bulb scale with a modified wet bulb temperature (see text). The B4SR is found by joining these points and reading off from the appropriate B4SR scale. The P4SR is then calculated as described in the text.

somewhat bizarre nature when converted to m/s. A wind speed of 75 ft/min is entered as 0.4 m/s. Other speeds are converted on the basis of 100 ft/min ≡ 0.5 m/s. There is really a separate nomogram for each wind speed, consisting of a scale for globe temperature (common to all) on the left and a wet bulb scale on the right.

Predicted four-hour sweat rate (P4SR)

Between these is a curved line giving the Basic four-hour sweat rate (B4SR) which is required in order to calculate the P4SR.

Calculation of modified wet bulb temperature

The nomogram is not necessarily entered with the observed wet bulb temperature. This requires modification as follows.:

    a. If the globe temperature, T_g, differs from the dry bulb temperature, T_a, the wet bulb temperature is increased by $0.4(T_g - T_a)$ degC.

    b. If the metabolic rate exceeds 63 W/m^2 (the rate appropriate for men sitting in chairs) the wet bulb temperature is increased by the amount indicated in the small inset chart (Fig. 9.5).

    c. For men wearing overalls the wet bulb temperature is increased by 1.0 degC.

These modifications are additive. For example, for men wearing overalls and working at 100 W/m^2 in the conditions $T_g = 45\ ^\circ C$, $T_a = 40\ ^\circ C$, $T_{wb} = 30\ ^\circ C$, the modified wet bulb temperature would be calculated as follows:

Wet bulb temperature	30.0 °C
Add $0.4(T_g - T_a)$	2.0 degC
Add for metabolic rate (from inset chart)	1.4 degC
Add for clothing	1.0 degC
Modified wet bulb temperature	34.4 °C

Calculation of B4SR

The nomogram is now entered with the globe temperature and modified wet bulb temperature, using the wet bulb scale appropriate for the wind speed. A line drawn between these two points intersects the appropriate B4SR line at the B4SR value. For wind speeds between 0.4 and 2.5 m/s the wet bulb temperature should be interpolated.

Calculation of P4SR

The P4SR is found by adding to the B4SR amounts which depend on the metabolic rate and clothing.

    a. Men sitting in shorts: P4SR = B4SR.

    b. Men working in shorts: P4SR = B4SR + $0.012(M - 63)$.

    c. Men sitting in overalls: P4SR = B4SR + 0.25.

    d. Men working in overalls: P4SR = B4SR + 0.25 + 0.017 $(M - 63)$.

9. *Indices of heat stress – steady state*

In a given environment the B4SR is different for each of these conditions because of the modification to the wet bulb temperature.

This is the most complicated of the indices considered here, but also probably the most precise. Its accuracy was established by complementary experiments conducted in Singapore (Macpherson, 1960) and other physiological observations have confirmed its superiority to other indices (Wyndham, 1954; Lind & Hellon, 1957; Brebner, Kerslake & Waddell, 1958b; Löfstedt, 1961; Wyndham *et al.* 1967a).

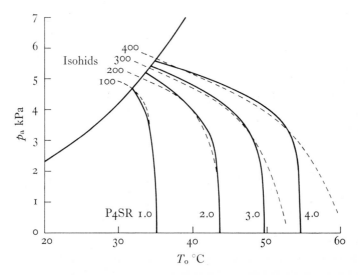

Fig. 9.6. Environments of equal P4SR plotted on the psychrometric axes. The subject is wearing shorts; $M = 100$ W/m², $V = 1.0$ m/s. The dashed lines are isohids for a nude subject who sweats according to the equation $S = H_p + 100$ $(T_s - 34.0)$. The isohids are for $H_p = 100$ W/m², $V = 1.0$ m/s. A sweat loss of 100 W/m² requires a total loss of 1.08 l in 4 h.

Lines of equal P4SR are shown on the psychrometric axes in Fig. 9.6 for a subject wearing shorts, $M = 100$ W/m², $V = 1.0$ m/s. The dashed lines are normal isohids for a subject who sweats according to the equation $S = H_p + 100(T_s - 34.0)$. If the arguments developed in Chapter 8 are accepted, these isohids should represent lines of equal stress for the nude subject. If one were attempting to predict the sweat loss, adjustment for hidromeiosis should be made, but since the P4SR is intended as an index of heat

Predicted four-hour sweat rate (P4SR)

stress rather than as a prediction of actual sweat rate the comparison
made in Fig. 9.6 seems more appropriate. Within the envelope of
conditions from which the P4SR was developed (see Fig. 9.7) the
agreement is close. A sweat loss of 100 W/m² requires a total loss
of about 1.08 l in 4 h. Some allowance should be made for the
initial warming-up period, so the numerical correspondence between the P4SR lines and the isohids is about what one would
expect.

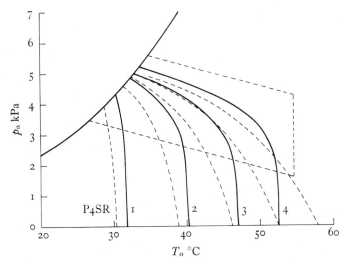

Fig. 9.7. Environments of equal P4SR plotted on the psychrometric axes. The
subject is wearing overalls; $M = 100$ W/m², $V = 1.0$ m/s. The curved dashed
lines are those of Fig. 9.4 for equal ITS under similar conditions of clothing,
work and air movement. The straight dashed lines indicate the envelope of
conditions used in the experiments upon which the P4SR was based.

The P4SR for subjects wearing shorts is adjusted for differences
in work rate by modifying the wet bulb temperature with which
the nomogram is entered, and by adding to the B4SR a quantity
of sweat about 10 per cent greater than would be required to
remove the additional metabolic heat, assuming full evaporation.
These adjustments have the general effect of shifting the curves
downwards and to the left with increasing heat production in the
same way as the isohids are shifted.

Adjustment for clothing is made in the same general way and has
the same effect. Clothing increases the index value under all con-

9. *Indices of heat stress – steady state*

ditions, whereas one might expect the opposite effect in hot dry environments with high air movement. Fig. 9.7 shows lines of equal P4SR for a clothed subject under the conditions of Fig. 9.4. The straight dashed lines indicate the highest and lowest wet and dry bulb temperatures used in the experiments on which the P4SR was based. Within this envelope the P4SR lines show a general correspondence with the ITS lines from Fig. 9.4, shown dashed.

The wet bulb globe temperature index (WBGT)

One of the most important sources of heat stress in army field operations is sunlight (cf. Table 3.4). The Effective Temperature scale (see below) in its original form made no allowance for additional radiant heat, and the adjustment favoured by Yaglou (1950) was regarded as too complicated for use in the field. Yaglou & Minard (1957) therefore devised a simpler substitute which would lead to much the same result. Three observations are required, the temperature of a standard black globe thermometer, T_g, the shaded bulb dry temperature, T_a, and the temperature of a wet bulb thermometer which is not artificially ventilated and is exposed to the ambient radiation, T'_{wb}. Details of the wet bulb thermometer and its use are given by Minard & O'Brien (1964).

The index consists of a simple weighting of the three temperatures.

$$\text{WGBT} = 0.7T'_{wb} + 0.2T_g + 0.1T_a.$$

An alternative formula can be used if the psychrometric wet bulb temperature (i.e. the reading of a forcibly ventilated wet bulb which is not exposed to radiation), T_{wb}, is available (Minard, 1964),

$$\text{WBGT} = 0.7T_{wb} + 0.3T_g.$$

The most striking aspect of this index is that no measurement of air movement is required. When the psychrometric wet bulb temperature is used, the only effect of air movement is on the globe temperature, and in the absence of a radiant heat load ($\overline{T}_r = T_a$), air movement would not affect the index value at all. The WBGT is therefore unlikely to be generally applicable to all types of hot environment, but in the open desert conditions for which it is intended the radiant heat load is normally high and air movement contributes to the index value through its effect on the globe thermometer.

Wet bulb globe temperature index (WBGT)

From a military point of view, the conditions of importance are those which will cause casualties through heat illness. The application of the WBGT is thus restricted to a band of conditions where the heat stress reaches a critical level and it is for such conditions that the WBGT is appropriate. For this band there must be some relation between temperature and humidity, and the fact that the same relation may not hold for other levels of heat stress is unimportant. Strictly one might expect the relation at the critical stress level to depend on the wind speed, but in the case of active, clothed subjects it may not be greatly influenced by wind speed within the range encountered in field operations.

The WBGT has proved of great value in reducing heat casualties in army training operations. It was introduced in 1956 at the Marine Corps Recruit Depot in South Carolina, with the recommendation that vigorous training be suspended when the WBGT reached 29.4 °C (recruits in the first three weeks of training) or 31.1 °C (all recruits). Although the weather was hotter than in the previous year, heat casualties fell to one third of the previous level, and less training time was lost (Minard, Belding & Kingston, 1957). The improvement has been maintained at stations where the WBGT is used (Minard & O'Brien, 1964), and the index is now applied widely in the control of military training programmes.

The success of the WBGT is due in part to its simplicity, which encourages its use and improves the reliability of observations. The wet bulb readings have been the source of some inaccuracy in the past, through misuse of the instrument, but fortunately the commonest errors (drying of the wick or positioning of the thermometer stem in the water reservoir) lead to falsely high readings and are therefore not dangerous.

When an index has proved so valuable in one context there is a temptation to apply it in others, and the administratively attractive notion of standardizing its use throughout all the armed services has inevitably been proposed. The WBGT was originally intended as a substitute for Yaglou's scale of Effective Temperature corrected for Radiation (ETR), restricted to the case of soldiers in the open. Yaglou and Minard did not intend it as a general alternative to the ETR and have not sought to defend its misuse. Minard, Kingston & van Liew (1960), in a study of conditions in an aircraft carrier, made only passing reference to the WBGT. It correlated poorly with measurements of physiological strain, and

9. *Indices of heat stress – steady state*

they did not consider it in their discussion of the merits of various indices of heat stress.

Lines of equal WBGT are shown on the psychrometric axes in Fig. 9.8. Since the wet bulb weighting is constant the lines are

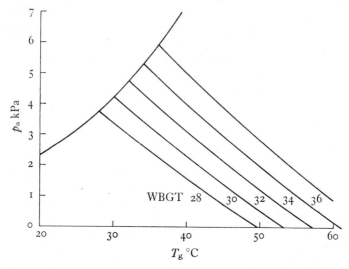

Fig. 9.8. Environments of equal WBGT plotted on the psychrometric axes. The WBGT does not make allowance for differences in clothing, work rate or air movement. It is intended for application only to troops training in the open.

nearly straight and parallel. Such curvature as they have is opposite to that of lines of equal ITS (Fig. 9.4) and equal P4SR (Fig. 9.7), and they do not show the steepening at lower stress levels characteristic of these indices, the HSI (Fig. 9.2) and the ETR (Fig. 9.11). Their relation to lines of equal ETR is considered on page 246.

Effective temperature (ET)

The Effective Temperature scale was developed by Yaglou and his associates as a sensory scale of warmth (Houghton & Yagloglou, 1923; 1924; Yaglou & Miller, 1925; Yaglou, 1927). It is based on instantaneous impressions of subjects moving back and forth from one conditioned room to another. At the time it was thought that these sensations would correspond with sensations after prolonged

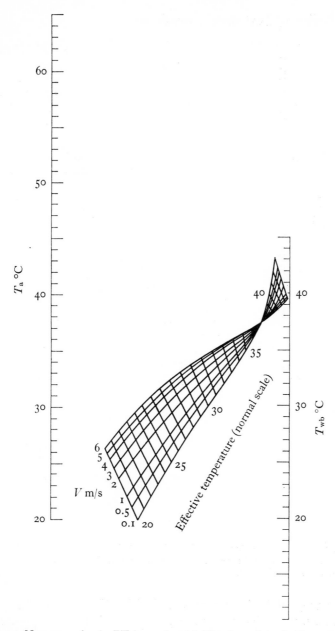

Fig. 9.9. Nomogram for the ET (normal scale). The dry and wet bulb temperatures are joined, and the ET read from the appropriate scale. The normal scale, shown here, applies to subjects wearing normal indoor clothing. Adjustments for radiation (CET and ETR) are described in the text.

9. Indices of heat stress – steady state

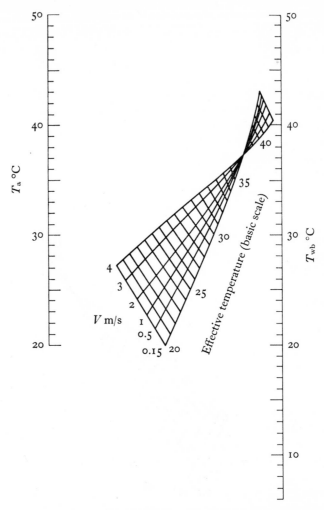

Fig. 9.10. Nomogram for the ET (basic scale). This scale is appropriate for subjects stripped to the waist. Adjustments for radiation (CET and ETR) are described in the text.

exposure, but this is now known to be false. However, the ET scale has been very widely used as a steady state index and appears to work quite well.

The Effective Temperature of an environment is the temperature of still, saturated air which would give rise to an equivalent sensation. The nomograms are for subjects wearing normal (1920s) American indoor clothing (normal scale) or stripped to the waist (basic scale). They are entered by joining the wet and dry bulb temperatures and reading the ET from the line appropriate for the prevailing air movement. In this original form no allowance is made for mean radiant temperature, which is assumed to equal the dry bulb temperature. Vernon & Warner (1932) proposed a radiation correction whereby globe temperature was used to enter the nomogram instead of dry bulb temperature. If the nomogram is entered in this way it can be argued that humidity is incorrectly represented, and that the wet bulb requires adjustment. Suppose that in a saturated environment at 30 °C (T_{wb} = 30 °C, p_a = 4.24 kPa) the radiant temperature is such that the globe thermometer reads 35 °C. The globe thermometer gives a correct indication of sensible heat stress, but the wet bulb temperature should be adjusted to correspond with the actual vapour pressure. If T_a were 35 °C and p_a 4.24 kPa the wet bulb temperature would be 31.05 °C, and this is the value with which the nomogram should be entered. The ET indicated by entering the nomogram in this way is known as the Effective Temperature including Radiation (ETR) (Yaglou, 1950). This adjustment is not very easy unless a psychrometric chart is available. A simpler procedure, approximately correct for the comfort zone, consists of adding to the wet bulb temperature 0.4 times the difference between globe and dry bulb temperature (Vernon & Warner, 1932). To apply such a correction implies great faith in the precision of the ET scale, and it has been found unnecessary in practice (Bedford, 1964, pp. 369–72). The Corrected Effective Temperature (CET) is obtained merely by entering the nomogram with the observed values of globe and wet bulb temperatures. Macpherson (1960, pp. 127–30) has confirmed that the CET makes appropriate allowance for radiation for both normal and basic scales.

In comfortable environments in the steady state humidity has little effect on sensation (Winslow, Herrington & Gagge, 1937*b*, *c*; Hick, Keeton & Glickman, 1938; Phelps & Vold, 1934; Glickman *et al.* 1950; Koch, Jennings & Humphreys, 1960) and if the ET is used as a steady state index it makes too much allowance for humidity in this part of the range (Minard, 1964). Yaglou (1947) proposed a modification to the scale for dry bulb temperatures between 21 and 26 °C, but considered that the original scale worked satisfactorily for the steady state at dry bulb temperatures above 32 °C. This last conclusion was based on physiological reactions rather than sensory impressions (McConnell & Houghten, 1923; Houghten & Yagloglou, 1924; Yaglou, 1927).

Smith (1952) examined the relation between the sweat loss and ET for men wearing overalls or shorts and with rates of heat production of 63 and 129 W/m². Wind speeds ranged from 0.05 to 4.0 m/s. He found that at each work rate the total sweat loss during a four-hour exposure could be predicted almost as well from the ET as from the P4SR scale. The basic ET scale was used for men wearing shorts and the normal scale when overalls were worn. When the observations were grouped according to ambient humidity (r.h. 10–30 per cent and 80–100 per cent) it was found that both scales overestimated the effect of humidity. At given ET the observed sweat rates were higher at low humidities. The prediction of sweat rate was poor at very low wind speeds (0.05–0.1 m/s). The lowest wind speed on the normal ET scale is 0.1 m/s and on the basic scale 0.15 m/s. These values are realistic practical minima, and it is hard to believe that accurate observations can be made at lower wind speeds. Any movement of the subject alters the heat exchange coefficients and also stirs the air around him. The subjects on whom the ET scales were based were walking about, while Smith's data were obtained from men who were either resting or step climbing. Even if the velocities in the undisturbed air stream are correctly quoted, these differences would lead one to expect discrepancies at low wind speeds.

Adam *et al.* (1952) examined the relation between ET and physiological responses (sweat rate, rectal temperature and pulse rate) in naval ratings stationed in Singapore. The experiments were modelled on those of McArdle *et al.* (1947), which Smith used for his examination. The conclusions were similar. When each state of clothing and work was considered separately the responses were highly correlated with ET, but it appeared that the wet bulb temperature was given undue weight and the adverse effect of very low air movement at high humidity was underestimated. Further work on highly acclimatized subjects confirmed this general picture (Adam *et al.* 1955).

In severe conditions approaching the limits of tolerance, wet bulb temperature becomes the predominant determinant of heat stress and the ET scale may underestimate its importance in this part of the range (Eichna *et al.* 1945; Wyndham *et al.* 1953; 1967*a*; Lind & Hellon, 1957; Macpherson, 1960, pp. 282–4).

The original ET scales refer to subjects engaged in the light activity of moving from room to room. For heavier work it is sometimes assumed that environments of equal ET will still be equivalent, and the scales have been applied without modification to such cases. Smith (1955) has proposed an extension to the ET scale which makes allowance for different rates of working. The stress

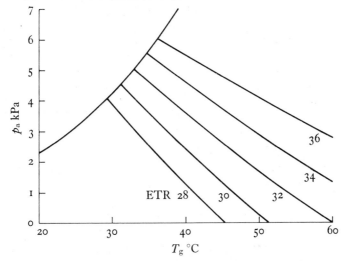

Fig. 9.11. Environments of equal ET, if the abscissa is entered with air temperature, or ETR if with Operative or globe temperature. The subject is clothed, $V = 1.0$ m/s. The ET scale does not make allowance for differences in work rate. If Smith's modification for work rate is used (Smith, 1955) the labels on the lines are changed, but each line still indicates environments of equal ET.

of a given environment and work rate can be expressed as the ET which would be associated with the same sweat rate at rest. However, the combinations of wet and dry bulb temperatures which are equivalent at any given work rate are derived from the original ET nomogram, and the device does not avoid the objection that environments which are equivalent at one work rate may not be equivalent at another.

Many of these criticisms of the ET scale are based on its inadequacy to predict physiological responses, in particular sweat rate. If hidromeiosis is a significant factor, or if humidity exerts a direct effect on sensation, sweat rate will not be a reliable indication of subjective distress. The alleged overemphasis of humidity in the lower and middle part of the range may not be as great as has been thought. However since the ET is based on initial impressions on entering an environment and is known to overemphasise the subjective effect of humidity in the steady state in comfortable environments, it seems likely that similar defects may be present in other parts of its range.

Lines of equal ET are shown in Fig. 9.11. These are derived from the normal scale, appropriate for clothed subjects. Vapour pressures have been calculated from the wet and dry bulb temperatures with which the nomogram was entered, so if the abscissa is globe temperature the figure shows lines of equal ETR, which do not necessarily correspond with CET. At any value of ET at a given wind speed the wet bulb weighting is constant. The lines therefore show a slight concavity similar to that of the WBGT lines in Fig. 9.8.

Comparison of ET and WBGT

Except in still air, the temperature of saturated air corresponding with a certain ET is higher than the ET, since the latter is the equivalent temperature of still saturated air. By its nature the WBGT in saturated air must be equal to the air temperature. ET lines and WBGT lines can therefore only coincide in still air, for which the WBGT is inappropriate. Although the labels on the lines must be different, the slopes of ET and WBGT lines are the same when the wet bulb weighting for the ET is 0.7. This occurs at ET 32 °C for a wind speed of 1.0 m/s. The temperature of saturated air at this wind speed and ET is 33.0 °C, and the line for WBGT 33 °C coincides with that for ET 32 °C.

Equivalences en séjour (ES)

This index was developed by Missenard (1948), who felt that the ET was unlikely to be appropriate for the steady state because it was based on sensations immediately after transfer from one environment to another. His observations are summarized in nomograms indicating environments which were judged equivalent either shortly after transfer (équivalences de passage) or after at least half an hour (équivalences en séjour). The French titles have

Ex.: $T_a = 35\ °C$, $T_{wb} = 30\ °C$, $V = 1\ m/s$, $ES = 32\ °C$.

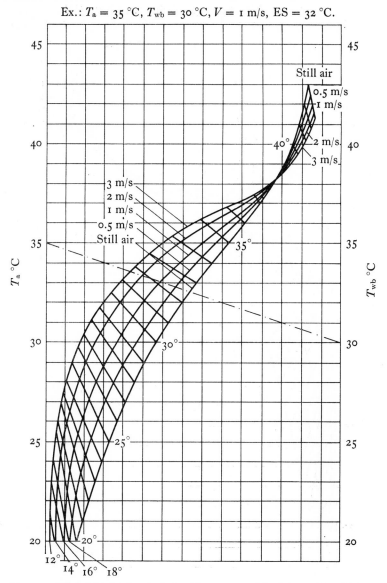

Fig. 9.12. ES nomogram for clothed subjects engaged in light activity. This is used in the same way as the ET nomograms.

9. *Indices of heat stress – steady state*

Ex.: $T_a = 38\ °C$, $T_{wb} = 32\ °C$, Still air, ES = 35 °C.

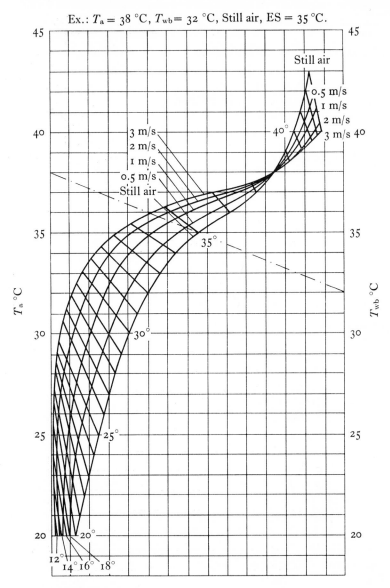

Fig. 9.13. ES nomogram for nude resting subjects. This is used in the same way as the ET nomograms.

been retained here in order to prevent confusion with Resultant Dry Temperature (Missenard, 1935), Resultant Temperature (Missenard, 1944), Equivalent Temperature (Dufton, 1936) and Equivalent Warmth (Bedford, 1936).

Nomograms for the Equivalences en séjour for nude and clothed subjects are reproduced in Figs. 9.12 and 9.13. They are used in the same way as the ET nomograms, and the index number for any environment is the temperature of still, saturated air which produces an equivalent sensation. Differences between subjects were such that the reliability of the ES was judged to be about ± 0.5 degC.

If these nomograms are accepted as indicating conditions which are subjectively equivalent in the steady state, they provide strong support for Missenard's belief that direct sensing of humidity contributes to the subjective assessment of environmental heat stress. At fixed air movement an equivalent sensation can be produced at various dry bulb temperatures. Sensible heat exchange must vary with dry bulb temperature, and cannot be the sole determinant of equivalence. The change in humidity which compensates for a change in dry bulb temperature will only affect total evaporation significantly if the wettedness is large, as it may be at high ES and high humidity. In other conditions the nomogram shows that humidity has an important effect on sensation.

It is of interest to see whether the nomogram is quantitatively consistent with the proposition that the subject bases his assessment of an environment on a combination of skin temperature and either skin relative humidity or wettedness. (In his analysis of the concept of equivalence, Missenard uses a term, μ, which he calls the 'humidité moyenne de la peau', but which corresponds mathematically to the wettedness.) The published material does not include measurements of skin temperature and sweat rate in warm environments, so it is impossible to examine this proposition directly. The approach which will be used here is to introduce a hypothetical subject whose skin temperature/sweat rate relation is $S = H_\mathrm{p} + 100(T_\mathrm{s} - 34.0)$. Skin heat loss is assumed to be 60 W/m², appropriate for heat balance in a resting subject, and on this basis the skin temperature, skin relative humidity and wettedness of a nude subject can be calculated.

Fig. 9.14 shows calculated values of T_s, ϕ_s and W for three values of ES at wind speeds of 0.5 and 2.0 m/s. Because of the assump-

9. *Indices of heat stress – steady state*

tions underlying the calculation and the way in which the nomogram is presented, the relation between T_s and ϕ_s at constant wind speed and ES must be nearly linear. However it is not inevitable that the lines for the same ES at different wind speeds should coincide. The fact that they do (nearly) is a consequence of the shape and position of the lines in the nomogram, and supports the proposition that skin relative humidity is sensed in some way.

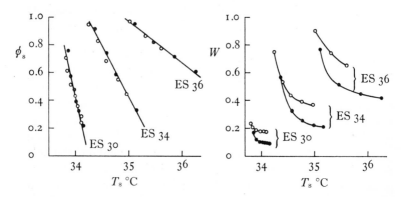

Fig. 9.14. Calculated values of T_s, ϕ_s and W for three values of ES (nude) at wind speeds of 0.5 (O) and 2.0 m/s (●). The assumptions underlying the calculation are described in the text, and would lead to linear relations between T_s and ϕ_s at each wind speed. However the coincidence of the lines for different wind speeds is a consequence of the form of the ES nomogram.

The lines for wettedness do not coincide at different wind speeds, as they would, approximately at least, if the sensation of humidity were due to limitation of evaporation from regions such as the forehead (p. 221).

The thermal sensation of clothed subjects continues to change for an hour after transfer from one environment to another (Glickman *et al.* 1947). Missenard's ES nomograms are based on sensations after at least half an hour, and show a smaller influence of humidity than do his observations shortly after entering the environment (équivalences de passage). It is possible that the influence of humidity would be even less when the subject had reached the steady state, but some residual effect remains even after several hours in any environment warmer than comfort (Koch, Jennings & Humphreys, 1960).

Lines of equal ES for clothed subjects engaged in light activity are shown in Fig. 9.15. They are two to three times as steep as the corresponding ET lines (Fig. 9.11).

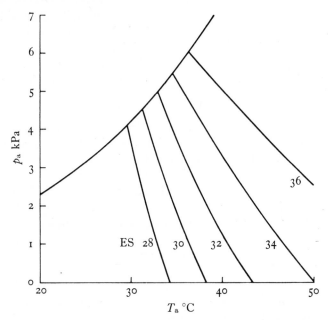

Fig. 9.15. Environments of equal ES (clothed) plotted on the psychrometric axes.

Comparison of heat stress indices

Choice of a heat stress index depends on the circumstances in which it will be applied. The sharpest knife is not the best one for opening envelopes. Of the two rationally based indices which have been considered here the ITS is the better. It is a little more complicated than the HSI, but this is offset by its more logical statement of heat stress. Its predictions of equivalence, unlike those of the HSI, accord with both theoretical expectation (Chapter 8) and practical experience (P4SR). Its structure allows for further development to cover other clothing assemblies.

The P4SR has proved accurate and valuable in practice, but because of its empirical nature its strict application is limited. It is comparable in complexity with the ITS, but makes no specific allowance for sunlight. Since the two indices lead to very similar results within the envelope of the P4SR experiments, the more flexible and comprehensive ITS is to be preferred.

The virtue of the WBGT lies in its simplicity. Only a straightforward weighting is required, which can easily be done electronically, and rugged portable instruments can be built to provide a direct reading. The application of the index is correspondingly restricted, not only to troops under training in the open, but to environments in which there is some danger of heat illness. The WBGT is based on the ET and may therefore not be the best objective guide to heat stress for most subjects. However, the men who matter are the ones who collapse, and there is very little information about the relation between environment and collapse for such subjects. The WBGT helps to identify critical conditions and can be used where more sophisticated indices would be misused or unused. In a permanent military base it is probable that an experienced medical orderly could rival the accuracy of the WBGT merely by sniffing the wind, but something more objective is required to support an expensive decision to stop training.

In the indices considered so far, equivalent environments are those which make an equal demand on the physiological defence mechanisms. They are not necessarily equally unpleasant, and where working efficiency or extra pay for adverse conditions is at issue the ET or ES scales may be more relevant. Of these the ES appears likely to be the better statement of steady state sensation, but the ET has been in widespread use for many years, and is understood and applied by people far removed from physiology. To change at this stage would undermine faith in all indices and would require more justification than at present exists.

Differences between subjects

None of the published indices of heat stress makes allowance for individual differences in shape, size and physiological responses. The point at issue is not that the strain in one subject may be greater than that in another in the same environment, but that a single family of equivalent environments, defined by the heat stress index, may not be appropriate for all subjects.

Differences in shape (including posture) and size affect the relation between air movement and heat exchange coefficients. The effects are not in general very large, and could be allowed for in the ITS by appropriate alteration of the coefficients, as is done for sunlight. However the accuracy of other components of the index would scarcely justify this in the case of convection and evaporation.

Physiological responses are a more important source of variation, and limit the precision of indices which do not make allowance for them. Fig. 9.16(a) shows skin isotherms for two subjects. These cross at an arbitrarily chosen reference environment, $T_a = 45\,°C$,

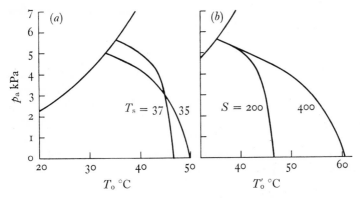

Fig. 9.16. The effect of physiological responses on environmental equivalence. $H_p = 100\ \text{W/m}^2$, $V = 1.0\ \text{m/s}$. (a) environments equivalent to $T_a = 45\,°C$, $p_a = 3.0\ \text{kPa}$, for subjects whose skin temperatures in this reference environment are 35 °C and 37 °C. (b) equivalent environments for two subjects having the same skin temperature (37.0 °C) but different sweat rates (200 and 400 W/m²).

$p_a = 3.0\ \text{kPa}$ (P4SR 2.4). The skin temperatures of 35 °C and 37 °C may represent a plausible range for these conditions. The greatest difference between the two families of equivalent environments is at $p_a = 5.0\ \text{kPa}$, where it is 7 degC. In saturated air the temperatures equivalent to the reference environment are 33 and 35 °C (P4SR 1.3 and 4.2 respectively).

Fig. 9.16(b) shows isotherms for $T_s = 37\,°C$ for sweat rates of 200 and 400 W/m². The wet skin lines coincide, but in dry air there is a difference of 14 degC in T_o. At very high humidities people with a low sweating capacity are at no disadvantage. So long as enough sweat is produced to wet the skin, subject responses do not influence families of equivalent environments. The important differences arise in hotter, drier environments, and here indices appropriate for acclimatized young men may be in serious error when applied to the general population.

9. Indices of heat stress – steady state

Nature of air movement

The extent of the zone of restricted evaporation depends on the nature as well as the magnitude of the air movement. In a turbulent omnidirectional air movement, ventilation of the skin is more uniform than in a wind, and the isohids approach the ideal case in which the transition from full evaporation to full wetness is abrupt (cf. Fig. 2.6). The importance of this is indicated by the differences between the ideal and cylinder isohids in Fig. 8.12. The greatest difference in T_0 can be expressed by the equation

$$\Delta T_0 = 0.18 S / h_0.$$

At $V = 1.0$ m/s, $S = 300$ W/m², the greatest difference in T_0 is 4 degC. The zone of restricted evaporation has not been thoroughly explored experimentally. Fortuitously, the inferences drawn from the behaviour of a uniformly sweating cylinder lead to satisfactory predictions of required sweat rate for nude subjects in a wind, but quantitative information about the effects of different types of air movement is lacking.

The ideal heat stress index

If the methods of analysis could be extended to include all environmental, sartorial and physiological factors it would be possible in principle to find the operating conditions of any subject under any conditions. This would not be the ideal heat stress index because it would be too complicated, but it would provide the material from which an index tailored to any specific requirements could be constructed. A good index is a map, not an aerial photograph, and its scale and content should be chosen in relation to its application. Its precision should match that of the data with which it will be used. Thus, 'military training' is not a very precise statement of physical activity, and although the energy costs of the various activities which comprise it may be known, it may be impossible to predict how troops will be employed on any particular day. With such uncertainty about heat production it would be inappropriate to apply a sophisticated index, and something very like the WBGT might well be the ideal.

10 HEAT STRESS AND TIME

If an initially cool subject is exposed to a hot environment he takes time to warm up. Although the environmental heat stress must be regarded as constant, the physiological strain increases as heat accumulates in the body. If some arbitrary upper limit of acceptable strain is set,the time taken to reach this level (tolerance time) will clearly depend on the environment, the characteristics of the subject and his thermal state on entry. The limit of strain need not be physical collapse of the subject; in many industrial situations a state in which performance begins to deteriorate may be of greater interest, and the definition of this state (as indeed that of the physiological limit) may itself involve time.

In very severe environments skin pain is usually the limiting factor (p. 257). Most subjects are able to breathe, albeit cautiously, in any environment tolerable for more than a few seconds, but a heedless gasp may set off an uncontrollable fit of coughing which must limit the exposure.

At more moderate rates of heat gain, the skin temperature does not reach painful levels, and tolerance may then be limited by the rise in tissue temperatures elsewhere. There appears to be a close relation between subjective distress and the total heat content of the body, and it is convenient to refer to such exposures as storage limited (p. 260). In most of the experimental work at these levels of stress the criterion of tolerance has been incipient collapse. Few investigators are prepared to carry exposures to the point of actual collapse, and the published tolerance times must necessarily depend on the methods used to define incipient collapse or the chosen levels of heart rate or core temperature at which exposures were terminated. Motivation is another important factor, since termination of the exposure at the urgent request of the subject is usually regarded as ethically mandatory. In the hot humid conditions most frequently used the subject's responses do not greatly affect cutaneous heat exchange.

255

10. *Heat stress and time*

Even under conditions in which heat balance can be attained the response characteristics of a given subject may change with time, and the strain may eventually reach an unacceptable level. Since the constitution of equivalent environments depends on the subject's responses, environments which are equivalent after, say, two hours exposure may not be so for the same subject after four hours.

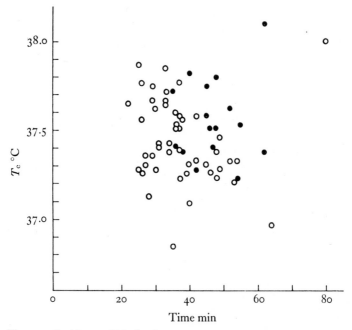

Fig. 10.1. Incidence of fainting in subjects who were passively tilted at different times after entering a hot environment. Filled symbols indicate the cases in which the subject fainted. The likelihood of fainting appears to depend on both the rectal temperature and the duration of exposure. (From Lind, Leithead & McNicol, 1968.)

Over periods of a few hours, dehydration is an important factor (Ferguson & Hertzman, 1958; Hertzman & Ferguson, 1959; Macpherson, 1960, p. 161). There is some evidence of a decline in circulatory performance within one hour. Lind, Leithead & McNicol (1968) have examined the deterioration in cardiovascular response to passive tilting during exposure at $T_a = 45 \,°C$, $p_a = 4.94 \, kPa$, $V = 0.5 \, m/s$. This environment can be tolerated by resting men for several hours. Subjects who were tilted during the first 35 min did not faint. After longer exposures some of the men fainted, and the results (Fig. 10.1) suggest that the likelihood of fainting may depend not only on the thermal state of the subject (as indicated by his rectal temperature) but also on the duration of exposure.

Postural fainting is usually a consequence of massive vasodilatation in skeletal muscle. Lind *et al.* (1968) found no evidence of this and suggested the splanchnic circulation as the site of the sudden fall in peripheral resistance. Blockley (1964) has described stepwise increases in rectal temperature in severely stressed subjects and has suggested intermittent closing of the splanchnic circulation as a possible cause. During the phases of rising rectal temperature the subjects were distressed and found it difficult or impossible to take water by mouth. When the rectal temperature was steady they felt better and drank readily.

Pain limited heat exposures

In very severe heat the duration of exposure may be limited because the skin becomes painfully hot. This is usually the critical factor for resting subjects when the tolerance time is less than 15 min (Webb, 1963). Buettner (1951 *b*) found that when small areas of skin were heated the pain became intolerable at a skin temperature of about 45 °C, although this depended a little on the rate of heating. He supplied heat from a radiant source at fixed rates, and compared his findings with the theoretical behaviour of a thick slab of material heated in this way. If the slab is initially at a uniform temperature the surface temperature will increase according to the equation

$$\Delta T_{\mathrm{s}} = \frac{2\dot{Q}\sqrt{t}}{\sqrt{(\pi.K.\rho.c)}}. \tag{1}$$

Here $\dot{Q}$ is the applied heat flux, t the time, K the thermal conductivity of the material, ρ its density and c its heat capacity per unit mass. The quantity $K.\rho.c$ is known as the thermal inertia of the material. If equation (1) were followed and the skin temperature at the end of the exposure were constant, the tolerance time, τ, would be related to $\dot{Q}$ by

$$\dot{Q} \propto \frac{1}{\sqrt{\tau}}. \tag{2}$$

10. *Heat stress and time*

Buettner found that in brief exposures (tolerance time less than 1 min) the skin temperature at the end of the exposure was higher the higher the heat flux. In longer exposures the value of $K.\rho.c$ appeared to change, possibly because the skin is not thick and homogeneous enough to behave as described by equation (1). Lipkin & Hardy (1954) suggest that local vascular responses to heat may be responsible. The observations have been confirmed by Stoll & Greene (1959). It may be incorrect to assume that there is no temperature gradient in the skin before heating begins. If a gradient is present the correct equation (Buettner, 1951 a) is

$$\Delta T_s = \left(\frac{\dot{Q}}{K}+\gamma\right)\frac{2\sqrt{(\alpha.t)}}{\sqrt{\pi}},$$

where α is the thermal diffusivity, $K/\rho.c$, and γ the initial temperature gradient in the material. At large values of $\dot{Q}$ (short tolerance times), γ is small compared with $\dot{Q}/K$ and may be neglected. For a tolerance time of 5 min, $\dot{Q}$ is about 1300 W/m². A representative value of K for skin is 0.63 W/m².°C (Webb, 1964). $\dot{Q}/K$ is then about 2000 °C/m, or 2 °C/mm. The temperature gradient, γ, is that obtaining in the upper layers of the skin before heating begins. The observations of Bazett (1951) and Reader & Whyte (1951) suggest a figure of about 0.5 °C/mm. This is not negligible compared with $\dot{Q}/K$, so the calculation of $K.\rho.c$ from equation (1) is invalid.

A further complication arises from the fact that skin transmits some radiation, and the heat is not all liberated at the surface. The radiant heating of diathermanous material has been treated in detail by Davis (1963).

Experiments on pain limited heat exposures are reported by Webb (1963). A pre-heated chamber was arranged on a track so that it could be rolled along to surround the subject. When this was done the radiant temperature around the subject changed suddenly, but the temperature of the surrounding air rose only slowly. It seems likely that in these brief exposures the predominant source of heat was radiation. Measurements showed that the tolerance limit was associated with skin temperatures of about 45 °C, consistent with previous work on small skin areas. The wall temperatures ranged from 100 °C to 290 °C, and since skin temperature changed only from 35 °C to 45 °C the radiant heat exchange during the exposure was almost constant. However total heat exchange would be considerably influenced by sweating, and it is hardly to be expected that the tolerance times would be well fitted by the theoretical equations described above. The effective radiant temperature (p. 58) is plotted against $1/\sqrt{\tau}$ in Fig. 10.2. The linear relation is consistent with equation (2), but probably results from a fortuitous cancellation of the two important factors which these equations neglect, changes in air temperature and sweating.

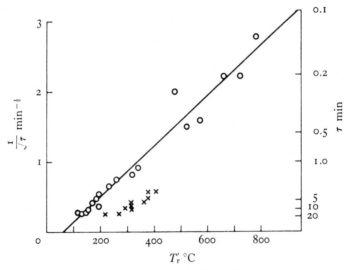

Fig. 10.2. Pain limited heat exposures. Ordinate, inverse of tolerance time; abscissa, effective mean radiant temperature (p. 58). Circles refer to experiments in which the radiant temperature was suddenly raised to the indicated value. Crosses refer to experiments in which the radiant temperature was increased throughout the exposure. The points are entered at the final effective mean radiant temperature. In both types of experiment the air temperature increased during the exposure. The subjects were nude. (Data of Webb, 1963.)

Webb also studied the effects of steadily increasing the wall temperature. If the flux of heat at the surface of a thick slab varies as t^n, the rise in surface temperature of the slab is proportional to $t^{(n+\frac{1}{2})}$ (Carslaw & Jaeger, 1959, p. 77). The value of the flux at the time when the surface temperature reaches a given level (e.g. the tolerance limit) is therefore proportional to $1/\sqrt{t}$, and equation (2) should hold, $\dot{Q}$ being the final flux. Webb's results for these 'ramp' experiments are plotted in this way in Fig. 10.2. For a given tolerance time the final wall temperature is higher in a

9-2

ramp experiment than when the wall temperature is constant throughout the exposure. This is to be expected, but the difference is not quantitatively consistent with the simple model of a slab receiving heat from a source at $\overline{T}'_r$.

When the subjects were clothed the tolerance times were greatly increased. The tolerance limit depends on pain from any region of skin, and is determined by the weakest link in the clothing protection. The knee-cap, for example, requires special attention because its blood flow appears to be rather small and in the sitting position the clothing may be drawn tightly over it.

Storage limited heat exposures

When the tolerance time is greater than about 15 min, skin pain and respiratory distress do not usually reach critical levels. The subject eventually becomes distressed and dizzy and may show personality changes. It is probable that syncope would follow, but this rarely occurs in the laboratory because exposures are terminated when it seems imminent. Criteria used by different workers vary widely, but within each investigation the results show that reproducibility is good. Differences between investigators are not as great as one might perhaps expect (Lind, 1964, pp. 87–8), and it may be that the final progression of symptoms and signs is so rapid that the clinical appraisal has little effect on the duration of exposure. However, Eichna *et al.* (1945), who drove their subjects with 'exhortations, blandishments and threats' despite chest pains, abdominal cramps and vomiting, noticed that under conditions classified as 'difficult' (see below) most of the men experienced a reaction like second wind during the second hour of work. At the end of four hours they were often judged to be in better condition than at the end of the first hour, when many observers might have considered that they had reached the limit of tolerance.

The condition of subjects near collapse is not identified with a critical level of skin or core temperature. Craig *et al.* (1954) estimated the total body heat content at the tolerance limit from measurements of skin and core temperature. Their subjects worked at three rates and wore various clothing assemblies. The end point was not complete exhaustion, but a state that the subject was willing to reach once a day, five days a week for five weeks. These 'voluntary tolerance times' ranged from 34 to 131 min, and the estimated body heat content at the tolerance limit was not significantly correlated with the tolerance time, i.e. it was substantially constant. The increase in stored heat during the exposure averaged 245 kJ/m². The observations of Blockley *et al.* (1954) can be treated in the same way. The increase in stored heat during their experiments was found by Craig *et al.* (1954) to be independent of the tolerance time, averaging 390 kJ/m². This larger figure may reflect the difference in tolerance criteria in the two series.

It is not at once obvious why the total body heat content should be so closely related to subjective distress or incipient collapse. Snellen (1966) found that sweating during exercise was proportional to the change in body heat content (measured calorimetrically), but sweating and cardiovascular responses are likely to be fully deployed before the tolerance limit is reached. A possible explanation runs as follows.

Suppose that at the tolerance limit or before it is reached the body conductance (related to peripheral blood flow) reaches a limiting value, C_{max}. Then the skin heat loss, H_s, is given by

$$H_s = C_{max}(T_c - T_s).$$

If the same core temperature were reached in the prescriptive zone, the skin heat loss, here designated H_z, would be regulated, as described on p. 132, at such a level that

$$H_z = b(T_c - T_{co}).$$

T_{co} is the core temperature at which H_z would be zero, about 36.5 °C. Suppose now that the subjective distress is proportional to the discrepancy between H_s, the actual rate of skin heat loss, and H_z, the preferred rate at the prevailing core temperature. This is an indication of the extent to which the normal physiological balance between core temperature and heat loss is overwhelmed by the environment.

$$\frac{(H_z - H_s)}{b} = \frac{(b - C_{max})}{b} T_c + \frac{C_{max}}{b} T_s - T_{co}.$$

T_{co} is constant, so the subjective distress will depend on a weighted mean of the skin and core temperatures. For fit young men, b is about 250 W/m².°C (p. 132) and C_{max} perhaps 80 W/m².°C (Fig. 7.22). These figures lead to a weighting of $0.68T_c + 0.32T_s$,

10. *Heat stress and time*

similar to the conventional weighting $(0.67T_c + 0.33T_s)$ used by Craig *et al.* (1954) to calculate body heat content. It is possible that in other subjects b and C_{max} may bear the same ratio to one another.

Heat stress and tolerance time

Whether the body heat content is directly concerned in determining the limit of tolerance or whether the relation between the two is fortuitous, it is worth examining the probable relation between tolerance time and environmental heat stress on the assumption that the body heat content at the tolerance limit is constant.

If the environment is very humid (dew point above 34 °C), water vapour will condense on the skin as the subject enters, and the skin may be assumed to be wet throughout the exposure. As the skin temperature rises sweat will be recruited, keeping the skin wet. It would be possible to treat the heat exchange in terms of the enthalpy difference between skin and air (Brunt, 1947) or by the methods of Chapter 4. In the present context there is some advantage in using wet and dry bulb temperatures (Houberechts, Lavenne & Patigny, 1958). The rate of heat loss from wet skin can be derived from equation (5), p. 80.

$$H_s = (h_0 + a.h_e)\left(T_s - \frac{(h_c + a.h_e)}{(h_0 + a.h_e)}T_{wb} - \frac{(h_0 - h_c)}{(h_0 + a.h_e)}T_a\right). \quad (3)$$

The equation is laid out in this way to show that H_s depends on a weighted mean of T_{wb} and T_a (assuming that $\bar{T}_r = T_a$). By analogy with Operative temperature, which deals with sensible heat exchange only, this weighted mean is here designated T_{oe}, and the overall heat exchange coefficient h_{oe} ($= h_0 + a.h_e$). Thus,

$$H_s = h_{oe}(T_s - T_{oe}). \quad (4)$$

The rate of heat storage is clearly $H_p - H_s$. If this were distributed evenly throughout the body, the quantity of heat, Q_t, stored after time t, would increase exponentially towards the quantity required for heat balance, Q_∞. When this quantity has been stored, the skin temperature will have risen to a level such that $H_s = H_p$. This need not be a physiologically attainable state.

$$Q_t = Q_\infty(1 - e^{-t/k}). \quad (5)$$

Here the time constant, k, is C/h_{oe}, where C is the heat capacity of the body per unit area. Machle & Hatch (1947) proposed a two-compartment model in which the ratio of heat stored in the core to that stored in the shell was constant. This leads to an equation of the same form, with a modified time constant.

The heat stored by a slab of material of finite thickness, initially at a uniform temperature, and exchanging heat with an environment at a different temperature, is described by the sum of a series of exponential terms (Carslaw & Jaeger, 1959, p. 121). The second and subsequent terms become negligible rather quickly. Fig. 10.3

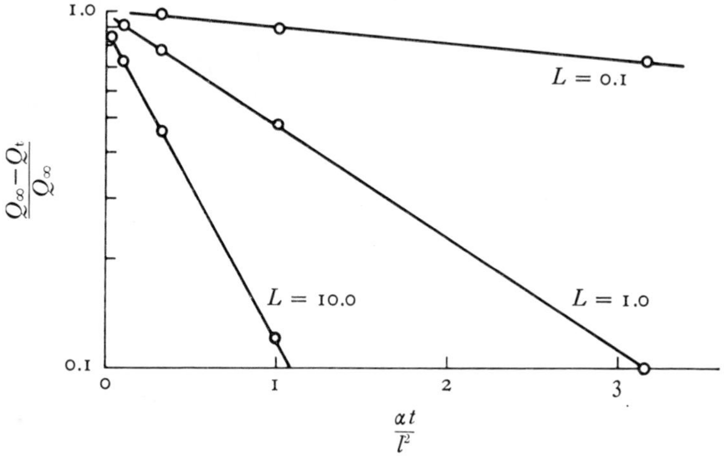

Fig. 10.3. Storage of heat in a slab of material, initially at a uniform temperature, exchanging heat with an environment at a different temperature. Ordinate, $(Q_\infty - Q_t)/Q_\infty$, where Q_t is the increase in heat content after time t and Q_∞ that after infinite time (logarithmic scale). Time is expressed as $\alpha . t/l^2$, where α is the thermal diffusivity of the material and l the half thickness of the slab. The parameter, L, is $l . h_{oe}/K$, where h_{oe} is the heat transfer coefficient at the surface of the slab and K the thermal conductivity of the material.

shows a semilog plot of the change in heat content with time (expressed as $\alpha . t/l^2$). A relation which is linear when plotted in this way can be described by equation (5). The parameter, L, is the half thickness of the slab, l, multiplied by the ratio of h_{oe} to the thermal conductivity of the material. It is difficult to guess analogous values for the human subject, but between the likely limits of $L = 10$ and $L = 0.1$, the increase in heat content can be adequately described by equation (5), with an appropriate value of k, so long as Q_t/Q_∞ is greater than about 0.1.

10. Heat stress and time

Since these three different examples all lead to substantially the same form of relation between heat storage and time, it may be supposed that the human body will behave in the same general way. Unfortunately it is impossible to calculate the appropriate time constant with any confidence. The observations of Wyndham *et al.* (1968) suggest that for resting subjects the value may be about 3 h.

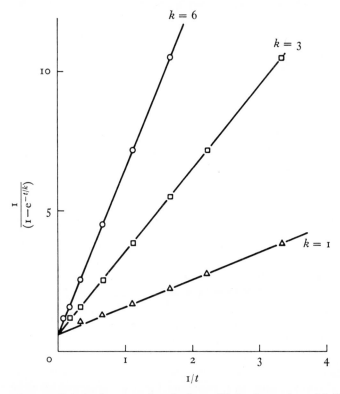

Fig. 10.4. Relation between $1/t$ and $1/(1 - e^{-t/k})$ for three values of k. The lines are described by equation (6) in the text.

For practical purposes a precise knowledge of the time constant is unnecessary. Although $(1 - e^{-t/k})$ is a markedly curvilinear function of t, the inverses of these quantities are linearly related over quite a large range. Fig. 10.4 shows examples for $k = 1, 3$ and 6 h. For values of t/k up to about 2, i.e. for durations up to

2 h for $k = 1$, 12 h for $k = 6$, the following equation is an adequate description.

$$\frac{1}{(1 - e^{-t/k})} = 0.98\frac{k}{t} + 0.6. \tag{6}$$

The lines in Fig. 10.4 are drawn according to this equation. Combining equations (5) and (6), and setting $k' = 0.98k$,

$$Q_\infty = Q_t\left(\frac{k'}{t} + 0.6\right).$$

Q_∞ is the quantity of stored heat which will raise the skin temperature to the level required for heat balance, $T_{oe} + H_p/h_{oe}$ (equation (4)). If the rise in skin temperature is proportional to Q,

$$T_{oe} + \frac{H_p}{h_{oe}} \propto Q_t\left(\frac{k'}{t} + 0.6\right). \tag{7}$$

When the tolerance limit is reached (at time τ), the stored heat has the constant value Q_τ. Tolerance time and environmental temperature will therefore be related by an equation of the general form

$$T_{oe} = \frac{A}{\tau} + B, \tag{8}$$

where A and B are parameters which depend on Q_τ, h_{oe} and H_p. Wyndham *et al.* (1968) used an equation of this type to describe their observations of the tolerance times of resting nude subjects in nearly saturated environments. The times extended up to 6 h. Other results, using different environments, clothing, work rates and tolerance criteria are also well fitted (see below).

Equivalent environments

If the skin becomes wet shortly after entering the environment, one would expect the tolerance time to be related to a weighted mean of the wet and dry bulb temperatures (equation (3)). The appropriate weightings for nude subjects are shown in Appendix 4. They are another way of describing the slopes of the wet skin isotherms developed in Chapter 8, and it is noteworthy that because the skin is wet the weighting is independent of the subject's physiological responses. The same weighting, which depends on air movement, posture and clothing, should apply for all subjects.

Lind (1964, p. 82) recommends a weighting of $0.85T_{wb} + 0.15T_a$ for all air movements, work rates and clothing. This weighting is

10. Heat stress and time

known as the 'W.D.' or 'Oxford' index, and its application reconciles the results of many different investigations quite well. Bidlot & Ledent (1947) prefer a weighting of $0.9T_{wb} + 0.1T_a$, and show that this is consistent with the observations of Eichna et al. (1945), Caplan (1943) and Caplan & Lindsay (1946).

The weighting which best fits the results of any experimental series in which T_{wb} and T_a are not confounded can be found by applying equation (8). Wyndham et al. (1968) found that the inverses of the tolerance times of their subjects in each environment were normally distributed, the variance being independent of the environment. This justifies the regression of $1/\tau$ on T_{wb} and T_a as a means of establishing the best weighting coefficients, provided that the relation between $1/\tau$ and T_{oe} turns out to be linear.

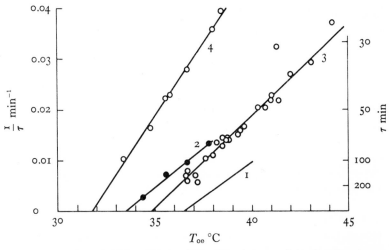

Fig. 10.5. Tolerance times of nude subjects. The inverse of the time in minutes is plotted against T_{oe}, a weighted mean of T_{wb} and T_a. 1, resting subjects (Wyndham et al. 1967c); 2, climbing at 220 m/h (Wyndham et al. 1967c); 3, resting subjects (Goldman, Green & Iampietro, 1965); 4, subjects working at 225 W/m² (Iampietro & Goldman, 1965).

Results for nude or nearly nude subjects are shown in Fig. 10.5. In the experiments of Wyndham et al. (1967c) the wet bulb depression was constant, so it is impossible to determine the weighting coefficients. These results are plotted against the wet bulb temperature, which must have been close to T_{oe} as the environments were nearly saturated. The men were either resting or step climbing at a rate equivalent to a vertical climb of 220 m/h. The line for an intermediate work rate falls between the two shown in Fig. 10.5. Goldman, Green & Iampietro (1965) and Iampietro & Goldman (1965) used less severe tolerance criteria than Wyndham *et al.* (1967c), and the tolerance times are shorter. Their choice of environments allows the best fit weighting coefficients to be determined, and these have been used in plotting the results. The individual tolerance times are not reported, and it has been necessary to use the inverse of the mean tolerance times rather than the more correct mean of the inverses of the tolerance times for each condition. The best fit weighting coefficients are $0.88T_{wb} + 0.12T_a$ for the resting subjects and $0.85T_{wb} + 0.15T_a$ for the working subjects. The mean air movements were 1.37 and 1.28 m/s respectively, and wet bulb weightings are smaller than the calculated value of 0.91 in Appendix 4. The heat transfer coefficients used by Houberechts, Lavenne & Patigny (1958) lead to a wet bulb weighting of 0.87.

The environments classed by Eichna *et al.* (1945) as impossible for sustained work, and of comparable severity, correspond with a weighting of $0.92T_{wb} + 0.08T_a$. Air movement was generated by the subjects marching at 1.3 m/s. Robinson, Turrell & Gerking (1945) used an air movement of 0.92 m/s. Their lines for E_p 300 (a high level of stress) for subjects wearing shorts and working at 218 W/m² and 151 W/m² indicate a weighting of $0.90T_{wb} + 0.10T_a$. Both these results are in good agreement with the values in Appendix 4. For resting subjects at E_p 300 Robinson *et al.* (1945) did not find a constant weighting. The mean for their line is about $0.87T_{wb} + 0.13T_a$.

The effect of air movement on the wet and dry bulb weighting factors is demonstrated indirectly by the results of Eichna *et al.* (1945). As was mentioned earlier, air movement was generated during the heat exposure by the subjects marching at 1.3 m/s. The air movement in the chamber is not stated and was probably low. Skin temperatures were measured at the end of the exposure, the subjects lying down. The authors comment that whereas rectal temperature, heart rate and work performance were affected about equally in different environments judged to be equivalent (and best fitted by the weighting $0.92T_{wb} + 0.08T_a$), skin temperature varied considerably, being lower in one of the most severe than in many of the milder environments (Fig. 10.6(*a*)). When the skin

10. *Heat stress and time*

temperature was measured, the air movement relative to the subject was low, and a lower wet bulb weighting would be more appropriate. If skin temperature is taken as the dependent variable the best weighting is $0.75T_{wb} + 0.25T_a$, consistent with a very low air movement. The observations are plotted against T_{oe} using this weighting in Fig. 10.6(*b*). Comparison with Fig. 10.6(*a*) shows the considerable effect of this change in weighting on the rank order of severity of these environments.

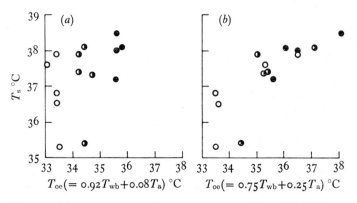

Fig. 10.6. Measurements of skin temperature by Eichna *et al.* (1945). (*a*) T_s plotted against $0.92\,T_{wb} + 0.08\,T_a$, the weighting which best described the general stress of the environments, and which is appropriate for the air movement of 1.3 m/s generated by the subjects. (*b*) T_s plotted against $0.75\,T_{wb} + 0.25\,T_a$, appropriate for a low air movement. The subjects were lying down when the measurements were made, and the air movement was probably low (see text). Environments which were classed as impossible, ●; difficult, ◑; easy, ○.

It might be thought that the skin temperature would not change fast enough for this explanation to be tenable. Fig. 10.7 shows changes in the skin temperature of the forehead of a subject sitting in a wind tunnel ($T_a = 48.5\,°C$, $T_{wb} = 30.0\,°C$, $V = 1.75$ m/s). He had previously exercised and was sweating profusely. The local heat exchange coefficients were altered by rotating the subject on a swivel chair so that he faced either directly into the wind or slightly downstream (100° of arc from the upstream position). This might change the local values of h_c and h_e about threefold. The response of skin temperature was rapid, the time constant being 18 s for rotation to the upstream position (left-hand diagram) and 25 s for rotation to the 100° position. Although the experiment does not reproduce the conditions used by Eichna *et al.* (1945) it shows that the skin temperature changes quite quickly. At a very low air movement one would expect a larger time constant than those in Fig. 10.7, and some of the variation of the results shown in Fig. 10.6 may be due to differences in the time taken to measure the skin temperatures.

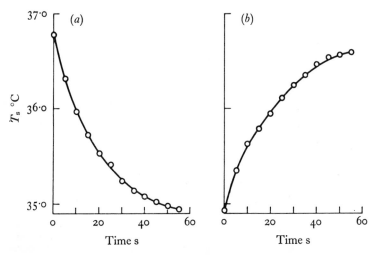

Fig. 10.7. Changes in skin temperature of the forehead of a subject sitting in a wind tunnel. $T_a = 48.5$ °C, $T_{wb} = 30.0$ °C, $V = 1.75$ m/s. He either faced directly into the wind or slightly downstream (100° of arc from the upstream position). (*a*) Changes after rotation from 100° position to upstream. (*b*) Changes after rotation from upstream position to 100°. Means of four experiments.

McConnell & Houghten (1923) measured voluntary tolerance times in still air. The subjects were stripped to the waist, wearing long trousers. The tolerance criteria are not stated in detail, and were probably not very precise, since the tolerance times of different subjects were frequently exactly equal. The results are shown in Fig. 10.8, where the inverses of the tolerance times are plotted against the best fit weighting of the environmental temperatures, $0.88T_{wb} + 0.12T_a$. Similar observations at higher air movements are reported by McConnell, Houghten & Yagloglou (1924). At an air movement of 2.0 m/s the best weighting is $0.93T_{wb} + 0.07T_a$. Results at 1.0 m/s are anomalous. The observations in saturated air fall on a straight line (plotting $1/\tau$ against T_{wb}), and the tolerance

10. *Heat stress and time*

time for saturated air at 42.8 °C can be interpolated. This is 26 min. At $T_a = 51.7$ °C, $T_{wb} = 42.8$ °C the observed tolerance time was 29 min. The results at 1.0 m/s have been plotted in the same way as those at 2.0 m/s in Fig. 10.8. The increased air movement does not seem to affect the tolerance times, which are but little shorter than those in still air.

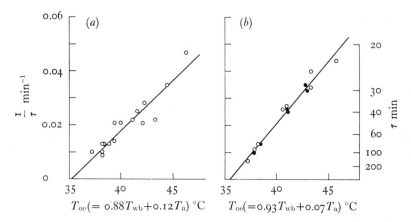

Fig. 10.8. Tolerance times of subjects stripped to the waist, wearing long trousers. (a) still air; $1/\tau$ plotted against $0.88\,T_{wb} + 0.12\,T_a$. (b) Air movement 1.0 m/s, ○, and 2.0 m/s, ●; $1/\tau$ plotted against $0.93\,T_{wb} + 0.07\,T_a$. (Data of McConnell & Houghten, 1923, and McConnell, Houghten & Yagloglou, 1924.)

Results for subjects wearing cotton overalls are shown in Fig. 10.9. The three sets of experiments used different subjects and were spaced some years apart. The air movements were not the same. In only one series (Provins *et al.* 1962) were the environments chosen in such a way that the weighting coefficients can be determined. In this case the best fit is given by $0.93T_{wb} + 0.07T_a$, and this weighting has been used for all the series shown in Fig. 10.9. This wet bulb weighting is high for an air movement of 0.25 m/s (the value in Appendix 4 for nude subjects is 0.82) and perhaps indicates that significant quantities of air were circulating beneath the overalls.

Blockley & Taylor (1949; 1950) measured tolerance times of subjects wearing long underwear and socks. Dry bulb temperatures up to 116 °C were used, and the ears were protected when necessary. The exposures were not terminated by skin pain. Ambient

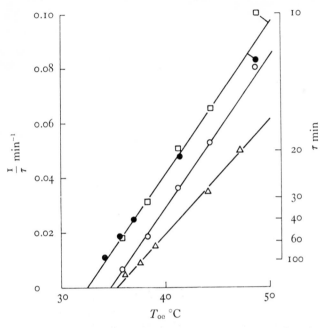

Fig. 10.9. Tolerance times of subjects wearing cotton overalls. $1/\tau$ plotted against $0.93\,T_{wb}+0.07\,T_a$. $\triangle$, resting, $V = 0.25$ m/s (Provins *et al.* 1962); $\bigcirc$, working at 81 W/m², $V = 0.75$ m/s (Bell *et al.* 1965); $\square$, working at 180 W/m², $V = 0.75$ m/s (Bell *et al.* 1965); $\bullet$, working at 180 W/m², $V = 1.0$ m/s (Bell & Walters, 1969).

vapour pressures ranged from 1.0 to 4.7 kPa. Results for two subjects are shown in Figs. 10.10 and 10.11. In the former the weighting $0.7T_{wb}+0.3T_a$ has been used for both subjects, and the results suggest that R.L. was better able to tolerate heat than S.F. If the subjects are considered separately, the best weighting for R.L. is $0.75T_{wb}+0.25T_a$, whereas for S.F. it is $0.64T_{wb}+0.36T_a$. The clothing probably did not permit circulation of air beneath it, and the wet bulb weightings may be low because of the reduced efficiency of sweat evaporation from clothing. The differences between subjects may be due to differences in fit of the clothing. It is also possible that the sweat rates were insufficient to maintain maximum evaporation in all conditions.

In Fig. 10.11 the wet and dry bulb temperatures have been weighted appropriately for each subject. This would be expected to reduce the scatter in each case, but not necessarily to bring the

10. *Heat stress and time*

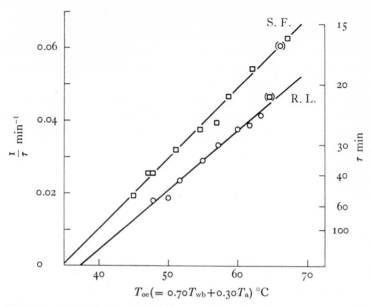

Fig. 10.10. Tolerance times of subjects wearing long underwear (Blockley & Taylor, 1950). $1/\tau$ plotted against $0.70T_{wb}+0.30T_a$. Results for each subject are plotted separately. R.L. appears to tolerate heat better than S.F. (but cf. Fig. 10.11).

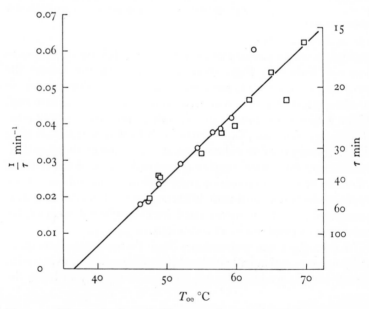

Fig. 10.11. The results shown in Fig. 10.10 plotted against the best fit weighing of T_{wb} and T_a for each subject (see text).

results for the two subjects closer together. It is now difficult to say which subject is the more 'tolerant to heat'.

These studies were extended to cover different clothing assemblies, and it was found that the tolerance time was closely related to the rate of heat storage, which could be calculated from knowledge of the clothing and environment. The product $\dot{Q}.\tau$ averaged 320 kJ/m² for exposures run to imminent collapse.† A minimum tolerance limit of 230 kJ/m² was suggested as safe for most subjects (Blockley *et al.* 1954).

Kaufman (1963) examined the problem of transient heat stress during re-entry of space vehicles. He calculated the sensible heat load from the environment and clothing, measuring the evaporative heat loss by weighing. His figure of 370 kJ/m² for the maximum tolerable heat storage is not very different from that of Blockley *et al.* (1954).

Precooling and preheating

In the experiments cited above subjects started from a state near comfort. If the tolerance limit is identified with a certain body heat content, the tolerance time will be greater if the subject starts from a condition cooler than comfort (Webb, 1959). Veghte & Webb (1961) have demonstrated this experimentally by precooling their subjects in air or water. The loss of stored heat calculated from the skin and rectal temperatures at the end of the cooling period does not agree quantitatively with the improvement in tolerance time if the conventional weighting of $0.67T_c + 0.33T_s$ is used. For cooling in water, however, it may be more correct to use a higher weighting for core temperature, and this leads to a closer agreement.

In practical situations it is more likely that men who have to go into a very hot environment will be somewhat warmer than comfort by the time they enter. Craig & Froelich (1968) preheated subjects before exhausting work at $T_a = 46\ ^\circ\mathrm{C}$, $T_{wb} = 23\ ^\circ\mathrm{C}$, minimal air movement. Endurance was reduced by preheating, but the times cannot be interpreted in simple terms because the work rate increased throughout the heat exposure. Wyndham *et al.* (1968) drew attention to the importance of preheating in emergency conditions in mines, where men would usually have been working in warm conditions before the emergency. They estimated the difference in body heat content at the start of the exposure to be

† If $\dot{Q}$ is in W/m², τ must be in seconds. $\dot{Q}.\tau$ is then in J/m².

10. *Heat stress and time*

about equal to one hour's heat production at rest, and proposed that the tolerance times of men who had been working would be about 1 h less than those of their experimental subjects, who had been cool before entering the hot environment.

Differences between subjects

Subjects may differ in the amount of heat storage they are able to tolerate, but provided that the skin is wet this should not affect the wet and dry bulb weighting which describes equivalent environments. This weighting is analogous to that of air and radiant temperatures in the expression of Operative temperature, also independent of the subject's responses but dependent on air movement. At a given air movement, T_o is the same for all subjects, as is T_{oe}. The characteristics of the subject become important when one attempts to compare Operative temperatures at different air movement, and a similar difficulty arises with T_{oe}. In equation (7) the term H_p/h_{oe} is the same for all subjects and is merely added to T_o. However, k' also depends on h_{oe}. For example, in the simplest case of even distribution of stored heat, $k = C/h_{oe}$. The effect of this is illustrated in Fig. 10.12.

Two subjects, having different values of Q_r, indicated by the dashed lines, are exposed to two environments, one at very high temperature and low air movement (large time constant), the other at lower temperature and higher air movement. The diagram has been arranged so that the tolerance time of subject A is the same in both environments. For him these environments are equivalent. They are not so for subject B. Thus when environments of different air movement are compared, equivalence based on the criterion of tolerance time must depend on the characteristics of the subject.

A level of heat storage lower than the average tolerance limit is sometimes chosen, either to allow for variations between subjects or to indicate a state associated with decrement in task performance, e.g. the minimum tolerance limit and minimum performance limit of Blockley *et al.* (1954). The effects of air movement will depend on the chosen value of heat storage, and environments of equal tolerance time based on one criterion will not necessarily be equivalent for another.

In practice the individuals who are least tolerant to heat are usually the most important. Wyndham (1962) has found that the rectal temperatures of men working in hot humid environments are not normally distributed. The distribution is skew, values much higher than the mean being commoner than a normal distribution would predict. By analysing this distribution he was able to predict the risk of heat stroke or heat collapse under given conditions of environment and work rate. He emphasizes that the conclusions

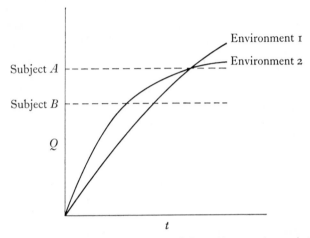

Fig. 10.12. The effect of characteristics of the subject on the equivalence of severe environments. Two subjects have different values of Q_T, indicated by the dashed lines. Heat is stored in the same way by both subjects in each environment, as indicated by the curves. For subject A the environments are equivalent. His tolerance times are the same. However subject B can tolerate the hotter environment with low air movement (1) longer than the lower temperature environment with high air movement (2). For him the environments are not equivalent.

are restricted to workmen in the South African gold mines. This work relates to exposures of several hours, but it seems probable that similar considerations may apply to tolerance times in very severe environments. Relatively small numbers of subjects have been used in tolerance time work, and the distribution of tolerance times has not been adequately explored. There is thus no sound basis on which the safe lower limits for exposure time can be predicted.

The wet and dry bulb weighting factors in equation (3) are based on the assumption that the skin is completely wet, and the experimental work cited earlier confirms that the values are approximately correct. The subjects have been fit young men, usually

10. *Heat stress and time*

acclimatized to heat, and the environments have usually been humid, circumstances which combine to make it likely that the condition of full wetness was met.

People whose maximum sweating capability is low are likely to be less able to tolerate heat than most others. If the environment is very humid they may be at no disadvantage, but at moderate humidities, particularly if the air movement is high, they may not produce sufficient sweat to keep the skin wet. If the skin is not wet, equation (3) cannot be applied. Evaporation is submaximal, and the methods of Chapter 8 must be used to estimate the skin heat loss. The portion of an isohid representing restricted evaporation cannot really be described by a weighting of wet and dry bulb temperatures, but in so far as it may be roughly represented in this way, the wet bulb weighting is smaller than for the wet portion of the isohid. The wet bulb weighting appropriate for normal young men (with wet skins) is therefore likely to be higher than that appropriate for the least tolerant subjects. The important effect of the weighting factors on the rank order of severity of the environments used by Eichna *et al.* (1945) has been indicated in Fig. 10.6. It would be almost impossible to determine by experiment the weighting factors appropriate for the most susceptible members of the general population. Emergency conditions do not occur with sufficient frequency to make Wyndham's (1962) epidemiological approach feasible in this context, and one of the commonest practical questions about heat stress remains unanswerable. 'What emergency conditions, for how long, will be harmless to all members of the general public who are fit enough to have made their way to this particular enclosure or vehicle?'

APPENDIX 1

Saturated water vapour pressure at various temperatures

The upper table gives values at 1 degC intervals from 0 °C to 79 °C, the lower table values at 0.1 degC intervals from 30 °C to 39.9 °C.

Temp °C	°C 0	1	2	3	4	5	6	7	8	9
0	0.61	0.66	0.71	0.76	0.81	0.87	0.93	1.00	1.07	1.15
10	1.23	1.31	1.40	1.50	1.60	1.70	1.82	1.94	2.06	2.20
20	2.34	2.49	2.64	2.81	2.98	3.17	3.36	3.56	3.78	4.00
30	4.24	4.49	4.75	5.03	5.32	5.62	5.94	6.27	6.62	6.99
40	7.37	7.78	8.20	8.64	9.10	9.58	10.08	10.61	11.15	11.73
50	12.33	12.96	13.61	14.29	15.00	15.73	16.50	17.31	18.14	19.01
60	19.91	20.85	21.83	22.84	23.90	25.00	26.14	27.32	28.55	29.82
70	31.15	32.51	33.94	35.42	36.95	38.54	40.18	41.87	43.63	45.46

	0	0.1	0.2	0.3	0.4	0.5	0.6	0.7	0.8	0.9
30	4.24	4.27	4.29	4.32	4.34	4.37	4.39	4.42	4.44	4.47
31	4.49	4.52	4.54	4.57	4.59	4.62	4.65	4.67	4.70	4.73
32	4.75	4.78	4.81	4.84	4.86	4.89	4.92	4.95	4.97	5.00
33	5.03	5.06	5.09	5.11	5.14	5.17	5.20	5.23	5.26	5.29
34	5.32	5.35	5.38	5.41	5.44	5.47	5.50	5.53	5.56	5.59
35	5.62	5.65	5.68	5.72	5.75	5.78	5.81	5.84	5.88	5.91
36	5.94	5.97	6.01	6.04	6.07	6.11	6.14	6.17	6.21	6.24
37	6.27	6.31	6.34	6.38	6.41	6.45	6.48	6.52	6.55	6.59
38	6.62	6.66	6.70	6.73	6.77	6.80	6.84	6.88	6.92	6.95
39	6.99	7.03	7.07	7.10	7.14	7.18	7.22	7.26	7.30	7.34

Saturated water vapour pressure, kPa.

APPENDIX 2

Wet bulb depression

Vapour pressure difference $(p_{wb} - p_a)$ in kPa at different wet bulb depressions, $(T_a - T_{wb})$. Example: to find the ambient water vapour pressure at $T_a = 35.0\ °C$, $T_{wb} = 31.6\ °C$. From Appendix 1, $p_{wb} = 4.65\ kPa$; from the table below the vapour pressure difference is 0.23 kPa (wet bulb depression $35.0 - 31.6 = 3.4\ degC$). Ambient water vapour pressure is therefore $4.65 - 0.23 = 4.42\ kPa$.

$(T_a - T_{wb})$ °C	0.0	0.2	0.4	0.6	0.8
0	0.00	0.01	0.03	0.04	0.05
1	0.07	0.08	0.09	0.11	0.12
2	0.13	0.15	0.16	0.17	0.19
3	0.20	0.21	0.23	0.24	0.26
4	0.27	0.28	0.30	0.31	0.32
5	0.34	0.35	0.36	0.38	0.39
6	0.40	0.42	0.43	0.44	0.46
7	0.47	0.49	0.50	0.51	0.53
8	0.54	0.55	0.57	0.58	0.59
9	0.61	0.62	0.64	0.65	0.66
10	0.68	0.69	0.70	0.72	0.73
11	0.74	0.76	0.77	0.79	0.80
12	0.81	0.83	0.84	0.85	0.87
13	0.88	0.90	0.91	0.92	0.94
14	0.95	0.96	0.98	0.99	1.01
15	1.02	1.03	1.05	1.06	1.08
16	1.09	1.10	1.12	1.13	1.14
17	1.16	1.17	1.19	1.20	1.21
18	1.23	1.24	1.26	1.27	1.28
19	1.30	1.31	1.32	1.34	1.35
20	1.37	1.38	1.40	1.41	1.42
21	1.44	1.45	1.47	1.48	1.49
22	1.51	1.52	1.54	1.55	1.56
23	1.58	1.59	1.61	1.62	1.63
24	1.65	1.66	1.68	1.69	1.71
25	1.72	1.73	1.75	1.76	1.78

APPENDIX 3

Psychrometric chart

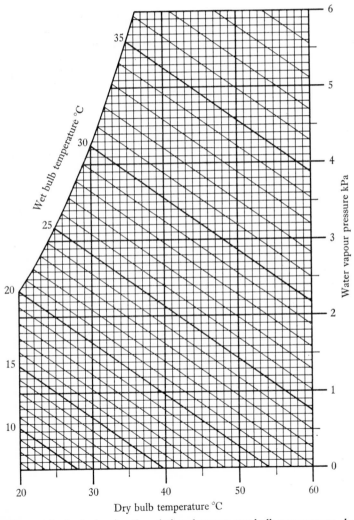

Psychrometric chart, showing the relations between wet bulb temperature, dry bulb temperature and ambient water vapour pressure. The chart is appropriate for standard sea level pressure (101.3 kPa).

APPENDIX 4

Heat exchange coefficients at various wind speeds

Values of heat exchange coefficients for nude subjects (sea level pressure). h_c and h_e are calculated as $8.3\sqrt{V}$ and $124\sqrt{V}$ respectively. The value of h_r is assumed to be 5.2 $W/m^2.°C$. The last two columns show the weightings of wet and dry bulb temperatures appropriate for a subject whose skin is completely wet.

V m/s	h_c W/m^2.°C	h_e W/m^2.kPa	h_o W/m^2.°C	Weighting T_{wb}	T_a
0.1	2.6	39	7.8	0.74	0.26
0.2	3.7	55	8.9	0.81	0.19
0.3	4.5	68	9.7	0.83	0.17
0.4	5.2	78	10.4	0.85	0.15
0.5	5.9	88	11.1	0.87	0.13
0.6	6.4	96	11.6	0.88	0.12
0.7	6.9	104	12.1	0.88	0.12
0.8	7.4	111	12.6	0.89	0.11
0.9	7.9	118	13.1	0.90	0.10
1.0	8.3	124	13.5	0.90	0.10
1.2	9.1	136	14.3	0.91	0.09
1.4	9.8	147	15.0	0.92	0.08
1.6	10.5	157	15.7	0.92	0.08
1.8	11.1	166	16.3	0.93	0.07
2.0	11.7	175	16.9	0.93	0.07
2.5	13.1	196	18.3	0.94	0.06
3.0	14.4	215	19.6	0.94	0.06
3.5	15.5	232	20.7	0.95	0.05
4.0	16.6	248	21.8	0.95	0.05
4.5	17.6	263	22.8	0.95	0.05
5.0	18.6	277	23.8	0.95	0.05

APPENDIX 5

Black body radiation at various temperatures

Values of $\sigma . T^4 (W/m^2)$ at different temperatures ($^\circ C$).

°C	0	1	2	3	4	5	6	7	8	9
0	315	320	325	329	334	339	344	349	353	358
10	364	369	374	379	385	390	396	401	407	413
20	418	424	430	436	442	448	453	459	465	472
30	478	484	491	497	504	511	517	524	531	537
40	544	551	558	565	573	580	587	595	602	610
50	618	625	633	641	649	656	664	673	681	689
60	697	706	715	723	732	741	750	758	767	776
70	785	794	804	813	823	832	842	851	861	871
80	881	891	901	911	922	932	943	952	963	974
90	985	996	1007	1018	1028	1041	1052	1063	1075	1086
100	1098	1110	1122	1134	1147	1158	1170	1183	1195	1208

APPENDIX 6

Body surface area from height and weight (from Best & Taylor, 1937)

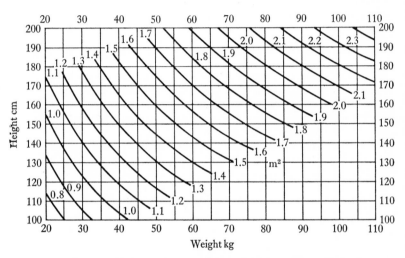

Chart for determining surface area in man in square metres. Example: weight 60 kg, height 170 cm, surface area = 1.70 m². (After DuBois.)

Relation of height and weight to surface area after DuBois. (© 1937 The Williams & Wilkins Co., Baltimore.)

Height (cm)	Weight (kg)																
	25	30	35	40	45	50	55	60	65	70	75	80	85	90	95	100	105
200							1.84	1.91	1.97	2.03	2.09	2.15	2.21	2.26	2.31	2.36	2.41
195						1.73	1.80	1.87	1.93	1.99	2.05	2.11	2.17	2.22	2.27	2.32	2.37
190				1.56	1.63	1.70	1.77	1.84	1.90	1.96	2.02	2.08	2.13	2.18	2.23	2.28	2.33
185				1.53	1.60	1.67	1.74	1.80	1.86	1.92	1.98	2.04	2.09	2.14	2.19	2.24	2.29
180				1.49	1.57	1.64	1.71	1.77	1.83	1.89	1.95	2.00	2.05	2.10	2.15	2.20	2.25
175	1.19	1.28	1.36	1.46	1.53	1.60	1.67	1.73	1.79	1.85	1.91	1.96	2.01	2.06	2.11	2.16	2.21
170	1.17	1.26	1.34	1.43	1.50	1.57	1.63	1.69	1.75	1.81	1.86	1.91	1.96	2.01	2.06	2.11	
165	1.14	1.23	1.31	1.40	1.47	1.54	1.60	1.66	1.72	1.78	1.83	1.88	1.93	1.98	2.03	2.07	
160	1.12	1.21	1.29	1.37	1.44	1.50	1.56	1.62	1.68	1.73	1.78	1.83	1.88	1.93	1.98		
155	1.09	1.18	1.26	1.33	1.40	1.46	1.52	1.58	1.64	1.69	1.74	1.79	1.84	1.89			
150	1.06	1.15	1.23	1.30	1.36	1.42	1.48	1.54	1.60	1.65	1.70	1.75	1.80				
145	1.03	1.12	1.20	1.27	1.33	1.39	1.45	1.51	1.56	1.61	1.66	1.71					
140	1.00	1.09	1.17	1.24	1.30	1.36	1.42	1.47	1.52	1.57							
135	0.97	1.06	1.14	1.20	1.26	1.32	1.38	1.43	1.48								
130	0.95	1.04	1.11	1.17	1.23	1.29	1.35	1.40									
125	0.93	1.01	1.08	1.14	1.20	1.26	1.31	1.36									
120	0.91	0.98	1.04	1.10	1.16	1.22	1.27										

APPENDIX 7

Conversion factors

Factors for converting into s.i. units some of the units which have been used in thermal physiology in the past.

Length	1 ft	= 0.305 m.
Area	1 ft²	= 0.0929 m².
Volume	1 ft³	= 0.0283 m³.
Velocity	1 m.p.h.	= 0.447 m/s.
	1 ft/min	= 0.00508 m/s,
	1 ft/s	= 0.305 m/s.
Mass	1 lb	= 0.454 kg.
Density	1 lb/ft³	= 16.0 kg/m³.
Pressure	1 lbf/ft²	= 47.88 Pa,
	1 lbf/in²	= 6.89 kPa,
	1 torr	= 0.133 kPa,
	1 mm Hg	= 0.133 kPa,
	1 mb	= 0.100 kPa.
Power	1 B/h	= 0.293 W,
	1 kcal/h	= 1.163 W,
	1 kpm/s	= 9.81 W,
	1 kpm/min	= 0.163 W.
Heat transfer	1 B/ft².h	= 3.15 W/m²,
	1 kcal/m².h	= 1.163 W/m².
Conductance	1 B/ft².h.°F	= 5.68 W/m².°C,
	1 kcal/m².h.°C	= 1.163 W/m².°C.
Conductivity	1 B/ft.h.°F	= 1.73 W/m.°C,
	1 cal/cm.s.°C	= 418.7 W/m.°C.

B = British thermal units
lbf = pounds force

284

APPENDIX 8

Temperature conversion table; Fahrenheit to Celsius

Temp. °F	0	1	2	3	4	5	6	7	8	9
50	10.00	10.56	11.11	11.67	12.22	12.78	13.33	13.89	14.44	15.00
60	15.56	16.11	16.67	17.22	17.78	18.33	18.89	19.44	20.00	20.56
70	21.11	21.67	22.22	22.78	23.33	23.89	24.44	25.00	25.56	26.11
80	26.67	27.22	27.78	28.33	28.89	29.44	30.00	30.56	31.11	31.67
90	32.22	32.78	33.33	33.89	34.44	35.00	35.56	36.11	36.67	37.72
100	37.78	38.33	38.89	39.44	40.00	40.56	41.11	41.67	42.22	42.78
110	43.33	43.89	44.44	45.00	45.56	46.11	46.67	47.22	47.78	48.33
120	48.89	49.44	50.00	50.56	51.11	51.67	52.22	52.78	53.33	53.89
130	54.44	55.00	55.56	56.11	56.67	57.22	57.78	58.33	58.89	59.44
140	60.00	60.56	61.11	61.67	62.22	62.78	63.33	63.89	64.44	65.00

Δ °F	0.1	0.2	0.3	0.4	0.5	0.6	0.7	0.8	0.9
Δ °C	0.06	0.11	0.17	0.22	0.28	0.33	0.39	0.44	0.50

APPENDIX 9

Important symbols

Other symbols are defined in the text near to where they occur.

A bar over a symbol indicates a mean value.

A dot over a symbol indicates rate of change of the quantity.

A_D	Body surface area calculated from the DuBois formula (m^2).
A_p	Projected area of the body for solar radiation (m^2).
A_r	Effective area of the body for radiant heat exchange (m^2).
a	Increase in saturated water vapour pressure per degree rise in temperature ($kPa/°C$).
C	Rate of heat exchange by convection per unit surface area (W/m^2).
C	Net conductance of the body per unit surface area ($= H_s/(T_c - T_s)$) ($W/m^2 . °C$).
C	Heat capacity of the body per unit surface area ($J/m^2 . °C$).
c	Heat capacity per unit mass ($J/kg . °C$).
D	Mass diffusivity (m/s^2).
E	Rate of heat exchange by evaporation per unit surface area (W/m^2).
E_{max}	Rate of heat loss by evaporation from wet skin ($= h_e(p_s - p_a)$) (W/m^2).
$E_{max.c}$	Rate of heat loss by evaporation from wet skin and clothing, per unit body surface area (W/m^2).
$E_{max.s}$	Rate of heat loss by evaporation from wet skin under clothing, per unit body surface area (W/m^2).
E_p	Index of physiological effect (Robinson) (dimensionless).
E_{req}	Rate of heat loss by evaporation required for heat balance, per unit body surface area ($W/m)^2$.
ES	Equivalences en séjour index (Missenard) (°C).
ET	Effective temperature index (Yaglou) (°C).
e	The base of natural logarithms, 2.718.
exp	Exponential ($\exp(x) = e^x$).
F	Rate of blood flow expressed as heat capacity flow ($W/°C$).
F_c	Configuration factor in radiant heat exchange (dimensionless).
F_ϵ	Emittance factor in radiant heat exchange (dimensionless).
$F_{\epsilon c}$	A factor in radiant heat exchange which includes emittance and configuration terms (dimensionless).
f	Function ($y = f(x)$ indicates that y and x are mathematically related in some way).
f_{pcl}	Permeation efficiency factor (dimensionless).
g	Acceleration due to gravity, 9.807 m/s^2.
H	Rate of heat transfer per unit surface area (W/m^2).
H_o	Rate of sensible heat exchange (W/m^2).

286

Important symbols

H_p	Rate of heat production (W/m²).
H_r	Effective radiant field ($= F.\sigma(T_r^4 - T_a^4)$) ($W$/m²).
H_s	Net rate of heat loss from the skin (W/m²).
H_σ	Rate of storage of heat in the body (W/m²).
HSI	Heat stress index (Belding & Hatch) (dimensionless).
h_c	Coefficient for heat exchange by convection (W/m². °C).
h_D	Mass transfer coefficient ((kg/m².s)/(kg/m³) $=$ m/s).
h_e	Coefficient for heat exchange by evaporation (W/m². kPa).
h_0	Combined coefficient for heat exchange by convection and radiation ($= h_c + h_r$) (W/m². °C).
h_0'	The value of h_0 when $V = 0.07$ m/s (W/m². °C).
h_{0c}	Combined coefficient for heat exchange by convection and radiation for the clothed subject ($= k_0 + k_v$) (W/m². °C).
h_{0e}	Combined coefficient for heat exchange by convection, radiation and evaporation from wet skin ($= h_0 + a.h_e$) (W/m². °C).
h_r	First power coefficient for heat exchange by radiation, including emittance and configuration terms (W/m². °C).
h_r'	First power coefficient for heat exchange by radiation (W/m². °C).
I_a	Ambient insulation ($= 1/h_0$) (°C. m²/W).
I_c	Clothing insulation (°C. m²/W).
I_s	Solar radiation flux (W/m²).
ITS	Index of thermal stress (Givoni) (g/h).
K	Rate of heat exchange by conduction per unit surface area (W/m²).
K	Thermal conductivity (W/m. °C).
K	Increase in sweat production (expressed in heat units) per degree rise in skin temperature (W/m². °C).
k	Time constant (s, min or h).
k	Thermal conductance (W/m². °C).
k_0	Effective conductance between skin and environment (T_0) for clothed subject (W/m². °C).
k_v	Effective conductance between skin and ventilating air (T_v) for clothed subject (W/m². °C).
L	Characteristic dimension of length (m).
log	Logarithm to base 10.
ln	Logarithm to base e.
M	Rate of metabolic energy production per unit body surface area (W/m²).
$\dot{m}$	Rate of mass transfer (evaporation) per unit area (g/m².s).
N_{Gr}	Grashof number ($= g.L^3.\theta/\nu^2$) (dimensionless).
N_{Le}	Lewis number ($= h_c/h_D.\rho.c$) (dimensionless).
N_{Nu}	Nusselt number ($= h_c.L/K$)/(dimensionless).
N_{Pr}	Prandtl number ($= \mu.c/K$) (dimensionless).
N_{Re}	Reynolds number ($= V.L/\nu$) (dimensionless).
N_{Sc}	Schmidt number ($= \nu/D$) (dimensionless).
N_{Sh}	Sherwood number ($= h_D.L/D$) (dimensionless).
p_a	Partial pressure of water vapour in the ambient air (kPa).
p_s	Saturated water vapour at skin temperature (kPa).
p_{wb}	Saturated water vapour pressure at wet bulb temperature (kPa).
P4SR	Predicted four hour sweat index rate (McArdle) (l).

Appendix 9

Q	Quantity of heat (J).
R	Rate of heat transfer by radiation per unit body surface area (W/m^2).
S	Rate of storage of heat in the body per unit body surface area (W/m^2).
S	Sweat rate expressed in equivalent heat units per unit body surface area (W/m^2).
S_o	Sweat rate in the absence of hidromeiosis (W/m^2).
S_{req}	Sweat rate required for heat balance (W/m^2).
T_a	Air temperature (= dry bulb temperature) (°C).
T_c	Body core temperature (°C).
T_g	Globe temperature (°C).
T_o	Operative temperature $\left(=\dfrac{h_c}{h_o}\,T_a+\dfrac{h_r}{h_o}\,\bar{T}_r\right)$ (°C).
T_o'	Standard Operative temperature $\left(=\dfrac{h_o}{h_o'}\,T_o+\left(1-\dfrac{h_o}{h_o'}\right)\,T_s\right)$ (°C).
T_{oc}	Clothed Operative temperature $\left(=\dfrac{k_o}{h_{oc}}\,T_o+\dfrac{k_v}{h_{oc}}\,T_v\right)$ (°C).
T_{oe}	Equivalent saturated air temperature $$\left(=\dfrac{(h_c+a.h_e)}{(h_o+a.h_e)}\,T_{wb}+\dfrac{h_r}{(h_o+a.h_e)}T_a\right)\ (°C).$$
$\bar{T}_r$	Mean radiant temperature (°C or °K).
$\bar{T}_r'$	Effective mean radiant temperature (°C).
T_s	Skin temperature (°C).
T_{so}	Skin temperature at which sweat rate equals rate of heat production (°C).
T_v	Temperature of ventilating air (within clothing) (°C).
t	Time (s, min or h).
V	Wind speed (m/s).
$\dot{V}$	Respiratory minute volume (l/min).
$\dot{V}_{O_2}$	Rate of oxygen consumption (l/min).
$\dot{V}_{O_2\,max}$	Maximum rate of oxygen consumption (l/min).
W	Rate of external working per unit body surface area (W/m^2).
W	Wettedness (= E/E_{max}) (dimensionless).
WBGT	Wet bulb globe temperature index (Yaglou) (°C).
α	Thermal diffusivity (= $K/\rho.c$) (m^2/s).
α_s	Absorptance for sunlight (dimensionless).
β	Coefficient of cubical expansion (dimensionless).
ϵ	Emittance (dimensionless).
η_s	Efficiency of sweating (=E/S) (dimensionless).
η_{sc}	Efficiency of sweating for clothed subject (= (rate of removal of heat from body by evaporation)/S) (dimensionless).
θ	Temperature difference (degC).
λ	Wavelength of thermal radiation (μm).
λ	Latent heat of vaporization (J/g).
λ_s	Heat removed from skin by evaporation (J/g).
μ	Dynamic viscosity ($N.s/m^2$).

Important symbols

ν Kinematic viscosity ($=\mu/\rho$) (m²/s).

ρ Density (kg/m³).

σ Stefan–Boltzmann constant, $5.67 \cdot 10^{-8}$ W/m². °K⁴.

τ Time taken to attain a given physiological state (tolerance time) (s, min or h).

ϕ_s Relative humidity of skin (defined by $E = h_e(\phi_s \cdot p_s - p_a)$) (dimensionless).

REFERENCES

Adam, J. M., Collins, J. A. G., Ellis, F. P., Irwin, J. O., Jack, J. W., John, R. T., Jones, R. M., Macpherson, R. K. & Weiner, J. S. (1955). Physiological responses to hot environments of young European men in the tropics, II & III. Further studies on the effects of exposure to varying levels of environmental stress. Rep. No. RNP 55/831, Med. Res. Council, London.

Adam, J. M., Ellis, F. P., Irwin, J. O., Thomson, M. L. & Weiner J. S. (1952). Physiological responses to hot environments of young European men in the tropics, I. A preliminary study to determine the effects of exposure for four hours twice weekly to varying combinations of air temperature, humidity and air movement. Rep. No. RNP 52/721, Med. Res. Council. London.

Adam, J. M., Fox, R. H., Grimby, G., Kidd, D. J. & Wolff, H. S. (1960). Acclimatization to heat and its rate of decay in Man. *J. Physiol., Lond.* **152**, 26–7 P.

Adams, R., Johnson, R. E. & Sargent, F. (1958). The osmotic pressure (freezing point) of human sweat in relation to its chemical composition. *Q. Jl exp. Physiol.* **43**, 241–57.

Adolph, E. F. (1946). The initiation of sweating in response to heat. *Am. J. Physiol.* **145**, 710–15.

(1949). Laboratory and field studies: Desert. In *Physiology of Heat Regulation and the Science of Clothing*, ed. Newburgh, L. H., pp. 330–8. Philadelphia: Saunders. Facsimile (1968), New York: Hafner.

Adolph, E. F. & Molnar, G. W. (1946). Exchange of heat and tolerances to cold in men exposed to outdoor weather. *Am. J. Physiol.* **146**, 507–37.

Aikas, E. & Piironen, P. (1963). Thermal exchanges of the human body in extreme heat. Tech. Rep. No. AMRL–TDR–63–86, Wright–Patterson Air Force Base, Ohio.

Albert, R. E. & Palmes, E. D. (1951). Evaporation rate patterns from small skin areas as measured by an infrared gas analyzer. *J. appl. Physiol.* **4**, 208–14.

Ambler, H. R. (undated). Notes on the climate of Nigeria with reference to personnel. Tropical Testing Establishment Rep. No. 355, Ministry of Supply, London.

American Society of Heating and Ventilating Engineers (1942). Measurement of the physical properties of the thermal environment: Report of the ASHVE Research Technical Advisory Committee on Instruments. *Heat. Pip. Air. Condit.* **14**, June, 382–5.

Aschoff, J. (1957). Warmeaustausch in einer Modellextremitat. *Pflügers Arch. ges. Physiol.* **264**, 260–71.

Astrand, I. (1960). Aerobic work capacity in men and women with special reference to age. *Acta physiol. scand.* **49**, Suppl. 169, 64–73.

References

Astrand, I, Astrand, P.-O. & Rodahl, K. (1959). Maximal heart rate during work in older men. *J. appl. Physiol.* **14**, 562–6.

Astrand, P.-O. (1952). *Experimental Studies of Physical Working Capacity in relation to Sex and Age.* Copenhagen: Munksgaard.

Banjeree, M. R., Elizondo, R. & Bullard, R. W. (1969). Reflex responses of human sweat glands to different rates of skin cooling. *J. appl. Physiol.* **26**, 787–92.

Banjeree, S. & Sen, R. (1955). Determination of the surface area of the body of Indians. *J. appl. Physiol.* **7**, 585–8.

Bar-Or, O., Lundegren, H. M. & Buskirk, E. R. (1969). Heat tolerance of exercising obese and lean women. *J. appl. Physiol.* **26**, 403–9.

Bass, D. E. (1963). Thermoregulatory and circulatory adjustments during acclimatization to heat in Man. In *Temperature, its Measurement and Control in Science and Industry*, Vol. 3, Pt 3, ed. Hardy, J. D., pp. 299–305. New York: Reinhold.

Bazett, H. C. (1951). Theory of reflex controls to explain regulation of body temperature at rest and during exercise. *J. appl. Physiol.* **4**, 245–62.

Bazett, H. C., Love, L., Newton, M., Eisenberg, L., Day, R. & Forster, R. (1949). Temperature changes in blood flowing in arteries and veins in Man. *J. appl. Physiol.* **1**, 3–19.

Bean, W. B. & Eichna, L. W. (1943). Performance in relation to environmental temperature; reactions of normal young men to simulated desert environment. *Fedn Proc. Fedn Am. Socs exp. Biol.* **2**, 144–58.

van Beaumont, W. & Bullard, R. W. (1963). Sweating: its rapid response to muscular work. *Science, N.Y.* **141**, 643–5.

Bedford, T. (1936). The warmth factor in comfort at work. *Rep. ind. Hlth Res. Bd., Lond.* No. 76. Cited by Bedford, 1964.

(1964). *Basic Principles of Ventilation and Heating.* 2nd ed. London: Lewis.

Bedford, T. & Warner, C. G. (1934). The globe thermometer in studies of heating and ventilation. *J. Hyg., Camb.* **34**, 458–73.

Belding, H. S. (1949). Protection against dry cold. In *Physiology of Heat Regulation and the Science of Clothing*, ed. Newburgh, L. H., pp. 351–70. Philadelphia: Saunders. Facsimile (1968) New York: Hafner.

Belding, H. S. & Hatch, T. F. (1955). Index for evaluating heat stress in terms of the resulting physiological strain. *Heat. Pip. Air Condit.* **27**, Aug. 129–36.

(1956). Index for evaluating heat stress in terms of resulting physiological strains. *Trans. Am. Soc. Heat. Refrig. Air-Condit. Engrs* **62**, 213–28.

(1963). Relation of skin temperature to acclimation and tolerance to heat. *Fedn Proc. Fedn Am. Socs exp. Biol.* **22**, 881–3.

Belding, H. S. & Hertig, B. A. (1962). Sweating and body temperatures following abrupt changes in environmental temperature. *J. appl. Physiol.* **17**, 103–6.

Belding, H. S., Hertig, B. A. & Kraning, K. K. (1966). Comparison of man's responses to pulsed and unpulsed environmental heat and exercise. *J. appl. Physiol.* **21**, 138–42.

Bell, C. R., Hellon, R. F., Hiorns, R. W., Nicol, P. B. & Provins, K. A. (1965). Safe exposure of men to severe heat. *J. appl. Physiol.* **20**, 288–92.

Bell, C. R. & Walters, J. D. (1969). Reactions of men working in hot and humid conditions. *J. appl. Physiol.* **27**, 684–6.

References

Benjamin, F. B. (1953). Sweating response to local heat application. *J. appl. Physiol.* **5**, 594–8.

Bennett, J. W. (1964). Thermal insulation of cattle coats. *Proc. Aust. Soc. Anim. Prod.* **5**, 160–6.

Bennett, J. W. & Hutchinson, J. C. D. (1964). Thermal insulation of short lengths of Merino fleece. *Aust. J. Agric. Res.* **15**, 427–45.

Benzinger, T. H. (1959). On physical heat regulation and the sense of temperature in Man. *Proc. natn Acad. Sci. U.S.A.* **45**, 645–59.

— (1961). The quantitative mechanism and the sensory receptor organ of human temperature control in warm environment. *Ann. intern. Med.* **54**, 685–99.

Best, C. H. & Taylor, N. B. (1937). *The Physiological Basis of Medical Practice.* London: Ballière, Tindall & Cox.

Bidlot, R. & Ledent, P. (1947). Travail dans les milieux à haute température. Que savons-nous des limites de température humainement supportables? Comm. No. 28, Institut d'Hygiene des Mines, Hasselt.

Blank, I. H. (1952). Factors which influence the water content of the stratum corneum. *J. invest. Derm.* **18**, 433–40.

— (1953). Further observations on factors which influence the water content of the stratum corneum. *J. invest. Derm.* **21**, 259–69.

Blockley, W. V. (1964). Changes in the boundary between neutral and stressful thermal conditions caused by respiratory protective equipment. Webb Associates, Yellow Springs, Ohio.

— (1965). A systematic study of the human sweat response to activity and environment in the compensable zone of thermal stress. NASA CR-65260, Washington, D.C.

— (1968). Dehydration and survivability in warm shelters. Final Rep. Contr. No. OCD–PS–64–201, Office of Civil Defense, Washington, D.C.

Blockley, W. V., McCutchan, J. W., Lyman, J. & Taylor, C. L. (1954). Human tolerance for high temperature aircraft environments. *J. Aviat. Med.* **25**, 515–22.

Blockley, W. V. & Taylor, C. L. (1949). Human tolerance limits for extreme heat. *Heat. Pip. Air Condit.* **21** May, 111–16.

— (1950). Studies in tolerance for extreme heat. Second summary report. A.F. Tech. Rep. No. 5831, Wright–Patterson Air Rorce Base, Ohio.

Boyd, E. (1935). *The Growth of the Surface Area of the Human Body.* Minneapolis: Univ. Minnesota Press.

Brebbia, D. R., Goldman, R. F. & Buskirk, E. R. (1957). Water vapor loss from the respiratory tract during outdoor exercise in the cold. *J. appl. Physiol.* **11**, 219–22.

Brebner, D. F., Clifford, J. M., Kerslake, D. McK., Nelms, J. D. & Waddell, J. L. (1961). Rapid acclimatization to heat in Man. FPRC Memo No. 177, Ministry of Defence, London.

Brebner, D. F. & Kerslake, D. McK (1961a). The effect of altering the skin temperature of the legs on the forearm sweat rate. *J. Physiol., Lond.* **157**, 363–9.

— (1961b). The effect of cyclical heating of the front of the trunk on the forearm sweat rate. *J. Physiol., Lond.* **158**, 144–53.

292

References

(1963). The effect of soaking the skin in water on the acclimatization produced by a subsequent heat exposure. *J. Physiol., Lond.* **166**, 13 P.

(1964). The time course of the decline in sweating produced by wetting the skin. *J. Physiol., Lond.* **175**, 295–302.

(1968). The effects of soaking the skin in water at various temperatures on the subsequent ability to sweat. *J. Physiol., Lond.* **194**, 1–11.

(1969). The relation between sweat rate and weight loss when sweat is dripping off the body. *J. Physiol., Lond.* **202**, 719–35.

Brebner, D. F., Kerslake, D. McK. & Waddell, J. L. (1956). The diffusion of water vapour through human skin. *J. Physiol., Lond.* **132**, 225–31.

(1958a). The relation between the coefficients for heat exchange by convection and by evaporation in Man. *J. Physiol., Lond.* **141**, 164–8.

(1958b). The effect of atmospheric humidity on the skin temperatures and sweat rates of resting men at two ambient temperatures. *J. Physiol., Lond.* **144**, 299–306.

Breckenridge, J. R. & Woodcock, A. H. (1950). Effects of wind on insulation of arctic clothing. Rep. No. EP 164, Office of the Quartermaster General, Natick.

Brown, A. C. & Brengelmann, G. I. (1970). The interaction of peripheral and central inputs in the temperature regulation system. In *Physiological and Behavioural Temperature Regulation*, ed. Hardy, J. D., Gagge, A. P. & Stolwijk, J. A. J., pp. 684–702. Springfield: Thomas.

Brown, W. K. & Sargent, F. (1965). Hidromeiosis. *Archs envir. Hlth* **11**, 442–53.

Brunauer, S., Emmett, P. H. & Teller, E. (1938). Adsorption of gases in multi-molecular layers. *J. Am. chem. Soc.* **60**, 309–19.

Brunt, D. (1947). Some physical aspects of the heat balance of the human body. *Proc. phys. Soc., Lond.* **59**, 713–26.

Buettner, K. (1934). Die Wärmeübertragung durch Leitung und Konvektion. Verdunstung und Strahlung in Bioklimatologie und Meteorologie. Veröffentlichungen des Preussischen Meteorologischen Instituts. Abhandlungen Bd x No. 5, Berlin. Cited by Aikas & Piironen, 1963.

(1935) *Biol. Zbl.* **55**, 356. Cited by Taylor & Buettner, 1953.

(1951a). Effects of extreme heat and cold on human skin. I. Analysis of temperature changes caused by different kinds of heat application. *J. appl. Physiol.* **3**, 691–702.

(1951b). Effects of extreme heat and cold on human skin. II. Surface temperature, pain and heat conductivity in experiments with radiant heat. *J. appl. Physiol.* **3**, 703–13.

(1953). Diffusion of water and water vapour through human skin. *J. Appl. Physiol.* **6**, 229–42.

(1959a). Diffusion of liquid water through human skin. *J. appl. Physiol.* **14**, 261–8.

(1959b). Diffusion of water vapour through small areas of human skin in a normal environment. *J. appl. Physiol.* **14**, 269–75.

Bullard, R. W., Banjeree, M. R., Chen, F., Elizondo, R. & MacIntyre, B. A. (1970). Skin temperature and thermoregulatory sweating: a control systems approach. In *Physiological and Behavioural Temperature Regulation*, ed. Hardy, J. D., Gagge, A. P. & Stolwijk, J. A. J., pp. 597–610. Springfield: Thomas.

References

Bullard, R. W. Banjeree, M. R. & MacIntyre, B. A. (1967). The role of the skin in negative feedback regulation of eccrine sweating. *Int. J. Bioclim. Biomet.* **11**, 93–104.

Bulmer, M. G. & Forwell, G. D. (1956). The concentration of sodium in thermal sweat. *J. Physiol., Lond.* **132**, 115–22.

Burch, G. E. (1945). Rate of water and heat loss from the respiratory tract of normal subjects in a subtropical climate. *Archs intern. Med.* **76**, 315–27.

Burch G. E. & Winsor, T. (1944). Rates of insensible perspiration (diffusion of water) locally through living and through dead human skin. *Archs intern. Med.* **74**, 437–44.

Burton, A. C. & Edholm, O. G. (1955). *Man in a Cold Environment.* London: Arnold. Facsimile (1970), New York: Hafner.

Burton, D. R. (1965). The thermal assessment of personal conditioning garments. Tech. Rep. No. 65263, Royal Aircraft Establishment, Farnborough.

Caplan, A. (1943). A critical analysis of collapse in underground workers on the Kolar goldfield. *Trans. Instn Min. Metall.* **53**, 95–180.

Caplan, A. & Lindsay, J. K. (1946). An experimental investigation of the effects of high temperatures on the efficiency of workers in deep mines. *Trans. Instn Min. Metall.* **56**, 163–210.

Carroll, D. P. & Visser, J. (1966). Direct measurement of convective heat loss from the human subject. *Rev. scient. Instrum.* **37**, 1174–80.

Carslaw, H. S. & Jaeger (1959). *Conduction of Heat in Solids.* 2nd ed. Oxford: Clarendon.

Chalmers, T. M. & Keele, C. A. (1951). Physiological significance of the sweat response to adrenaline in Man. *J. Physiol., Lond.* **114**, 510–14.

Chrenko, F. A. & Pugh, L. G. C. E. (1961). The contribution of solar radiation to the thermal environment of Man in Antarctica. *Proc. R. Soc.* B **155**, 243–65.

Christie, R. V. & Loomis, A. L. (1932). The pressure of aqueous vapour in the alveolar air. *J. Physiol., Lond.* **77**, 35–48.

Clifford, J. C. (1966). Regional heat loss in Man in a state of thermal comfort. FPRC Rep. No. 1261, Ministry of Defence, London.

Clifford, J. C., Kerslake, D. McK. & Waddell, J. L. (1959). The effect of wind speed on maximum evaporative capacity in Man. *J. Physiol., Lond.* **147**, 253–9.

Cole, P. (1954). Recordings of respiratory air temperature. *J. Lar. Otol.* **68**, 295–307.

Colin, J. & Houdas, Y. (1965). Initiation of sweating in Man after abrupt rise in environmental temperature. *J. appl. Physiol.* **20**, 984–90.

— (1967). Experimental determination of coefficient of heat exchanges by convection of human body. *J. appl. Physiol.* **22**, 31–8.

Collins, K. J., Crockford, G. W. & Weiner, J. S. (1965). Sweat-gland training by drugs and thermal stress. *Archs envir. Hlth* **11**, 407–20.

Collins, K. J., Sargent, F. & Weiner, J. S. (1959a). Excitation and depression of eccrine sweat glands by acetylcholine, acetyl-β-methylcholine and adrenaline. *J. Physiol., Lond.* **148**, 592–614.

— (1959b). The effect of arterial occlusion on sweat-gland responses in the human forearm. *J. Physiol., Lond.* **148**, 615–24.

References

Collins, K. J. & Weiner, J. S. (1961). Axon reflex sweating. *Clin. Sci.* **21**, 333–44.

(1962*a*). Observations on arm-bag suppression of sweating and its relationship to thermal sweat-gland fatigue. *J. Physiol., Lond.* **161**, 538–56.

(1962*b*). The control and failure of sweating in Man. In *Biometeorology II*, ed. Tromp, S. W., pp. 280–5. London: Pergamon.

(1968). Endocrinological aspects of exposure to high environmental temperatures. *Physiol. Rev.* **48**, 785–839.

Coon, J. M. & Rothman, S. (1941). The sweat response to drugs with nicotine-like action. *J. Pharmac. exp. Ther.* **73**, 1–11.

Craig, F. N. & Froehlich, H. L. (1968). Endurance of preheated men in exhausting work. *J. appl. Physiol.* **24**, 636–9.

Craig, F. N., Garren, H. W., Frankel, H. & Blevins, W. B. (1954). Heat load and voluntary tolerance time. *J. appl. Physiol.* **6**, 634–44.

Crank, J. (1956). *The Mathematics of Diffusion*. Oxford: O.U.P.

Crockford, G. W. & Goudge, J. (1970). Dynamic insulation: an equation relating airflow, assembly thickness and conductance. *Proc. R. Soc. Med.* **63**, 1012–13.

Custance, A. C. (1965). Use of small skin surface areas for whole body sweating assessment. *Can. J. Phys. Pharmac.* **43**, 971–7.

Custance, A. C., Heath, C. & Cattroll, S. W. (1970). Insensible sweating; an intermediate stage between insensible perspiration and active sweating. DREO Rep. No. 610, Defence Research Establishment, Ottawa.

Davis, T. P. (1963). The heating of skin by radiant energy. In *Temperature, its Measurement and Control in Science and Industry*, Vol. 3, Pt 3, ed. Hardy, J. D., pp. 149–69. New York: Reinhold.

Dirnhuber, P. & Tregear, R. T. (1960). Equilibration between water vapour and human skin. *J. Physiol., Lond.* **152**, 58–59P.

Dobson, R. L. & Lobitz, W. C. (1966). The method of acclimatization in the human eccrine sweat gland. In *Human Adaptability and its Methodology*, ed. Yoshimura, H. & Weiner, J. S., pp. 158–65. Tokyo: Jap. Soc. prom. Sci.

DuBois, D. & DuBois, E. F. (1915). The measurement of the surface area of Man. *Archs intern. Med.* **15**, 868–81.

DuBois, E. F. (1924). *Basal Metabolism in Health and Disease*. Philadelphia: Lea & Febiger.

(1936). Mechanism of heat loss in health and disease. *Trans. Ass. Am. Physns.* **51**, 252–9.

Dufton, A. F. (1936). The equivalent temperature of a warmed room. *J. Instn Heat. Vent. Engrs* **4**, 227–9.

Dunham, W., Holling, H. E., Ladell, W. S. S., McArdle, B., Scott, J. W., Thomson, M. L. & Weiner, J. S. (1946). The effects of air movement in severe heat. Rep. No. RNP 46/316, Med. Res. Council, London.

Edholm, O. G., Fox, R. H. & Macpherson, R. K. (1957). Vasomotor control of the cutaneous blood vessels in the human forearm. *J. Physiol., Lond.* **139**, 455–65.

Eichna, L. W., Ashe, W. F., Bean, W. B. & Shelley, W. B. (1945). The upper limits of environmental heat and humidity tolerated by acclimatized men working in hot environments. *J. ind. Hyg. Toxicol.* **27**, 59–84.

References

Eichna, L. W., Park, C. R., Nelson, N., Horvath, S. M. & Palmes, E. D. (1950). Thermal regulation during acclimatization in a hot dry (desert type) environment. *Am. J. Physiol.* **163**, 585–97.

Ernsting, J. (1965). Respiration and anoxia. In *Textbook of Aviation Physiology*, ed. Gillies, J. A., pp. 209–63. London: Pergamon.

Fanger, P. O. (1967). Calculation of thermal comfort: Introduction of a basic comfort equation. *Trans. Am. Soc. Heat. Refrig. Air-Condit. Engrs* **73**, Pt 2, III. 4. 1–16.

—— (1970). *Thermal Comfort.* Copenhagen: Danish Technical Press.

Felsher, Z. (1944). Hereditary ectodermal dysplasia. *Arch. Derm. Syph.* **49**, 410–14.

Ferguson, I. D. & Hertzman, A. B. (1958). Regulation of body temperature during continuous exposure to heat. Rep. No. TR57–727, Wright Air Development Center, Ohio.

Ferguson, I. D., Hertzman, A. B., Rampone, A. J. & Christensen, M. L. (1956). Magnitudes, variability and reliability of regional sweating rates in humans at constant ambient temperatures. Rep. No. 56–38, Wright Air Development Center, Ohio.

Fishenden, M. & Saunders, O. A. (1950). *An Introduction to Heat Transfer.* Oxford: Clarendon.

Fitzgerald, L. R. (1957). Cutaneous respiration in Man. *Physiol. Rev.* **37**, 325–36.

Flink, C. H. (1960). *Heating, Ventilating and Air Conditioning Guide*, Vol. 38, pp. 14–15. New York: Am. Soc. Heat. Refrig. Air-Condit. Engrs.

Folk, G. E. & Peary, R. E. (1951). Water penetration into the foot. Rep. No. 181, Environmental Protection Section, Office of the Quartermaster General, Natick.

Fonseca, G. F., Breckenridge, J. R. & Woodcock, A. H. (1959). Wind penetration through fabric systems. Tech. Rep. No. EP–104, Office of the Quartermaster General, Natick.

Foster, K. G. (1961). Relation between the colligative properties and chemical composition of sweat. *J. Physiol., Lond.* **155**, 490–7.

Fox, R. H., Crockford, G. W. & Löfstedt, B. (1968). A thermoregulatory function test. *J. appl. Physiol.* **24**, 391–400.

Fox, R. H. & Edholm, O. G. (1963). Peripheral circulation in Man: Nervous control of the cutaneous circulation. *Br. med. Bull.* **19**, 110–14.

Fox, R. H., Goldsmith, R., Hampton, I. F. G. & Lewis, H. E. (1964). The nature of the increase in sweating capacity produced by heat acclimatization. *J. Physiol., Lond.* **171**, 368–76.

Fox, R. H., Goldsmith, R., Kidd, D. J. & Lewis, H. E. (1963). Acclimatization to heat in Man by controlled elevation of body temperature. *J. Physiol., Lond.* **166**, 530–47.

Fox, R. H. & Hilton, S. M. (1958). Bradykinin formation in human skin as a factor in heat vasodilation. *J. Physiol., Lond.* **142**, 219–32.

Fox, R. H., Löfstedt, B. E., Woodward, P. M., Eriksson, E. & Werkstrom, B. (1969). Comparison of thermoregulatory function in men and women. *J. appl. Physiol.* **26**, 444–53.

Froese, G. & Burton, A. C. (1957). Heat losses from the human head. *J. appl. Physiol.* **10**, 235–41.

Gagge, A. P. (1936). The linearity criterion as applied to partitional calorimetry. *Am. J. Physiol.* **116**, 656–68.

(1937). A new physiological variable associated with sensible and insensible perspiration. *Am. J. Physiol.* **120**, 277–87.

(1940). Standard Operative temperature. A generalized temperature scale applicable to direct and partitional calorimetry. *Am. J. Physiol.* **311**, 93–103.

(1941). Standard operative temperature, a single measure of the combined effect of radiant temperature, of ambient air temperature and of air movement on the human body. In *Temperature, its Measurement and Control in Science and Industry*, pp. 544–52. New York: Reinhold.

(1970). Effective radiant flux, an independent variable that describes thermal radiation on Man. In *Physiological and Behavioural Temperature Regulation*, ed. Hardy, J. D., Gagge, A. P. & Stolwijk, J. A. J., pp. 34–45. Springfield: Thomas.

Gagge, A. P., Burton, A. C. & Bazett, H. C. (1941). A practical system of units for the description of the heat exchange of Man with his thermal environment. *Science, N.Y.* **94**, 428–30.

Gagge, A. P., Graichen, H., Stolwijk, J. A. J., Rapp, G. M. & Hardy, J. D. (1968). ASHRAE-sponsored research project RP-41 produces R-meter. ASHRAE J. **10**, June, 77–81.

Gagge, A. P. & Hardy, J. D. (1967). Thermal radiation exchange of the human body by partitional calorimetry. *J. appl. Physiol.* **23**, 248–58.

Gagge, A. P., Herrington, L. P. & Winslow, C.-E. A. (1937). Thermal interchanges between the human body and its atmospheric environment. *Am. J. Hyg.* **26**, 84–102.

Gagge, A. P., Rapp, G. M. & Hardy, J. D. (1967). The effective radiant field and operative temperature necessary for comfort with radiant heating. *Trans. Am. Soc. Heat. Refrig. Air-Condit. Engrs* **73**, Pt 1, 1. 2, 1–8.

Gagge, A. P., Stolwijk, J. A. J. & Hardy, J. D. (1965). A novel approach to measurement of Man's heat exchange with a complex radiant environment. *Aerospace Med.* **36**, 431–5.

Gagge, A. P., Stolwijk, J. A. J. & Saltin, B. (1969). Comfort and thermal sensations and associated physiological responses during exercise at various ambient temperatures. *Envir. Res* **2**, 209–29.

Gates, D. M. (1962). *Energy Exchange in the Biosphere*. New York: Harper & Ross.

Gerking, S. D. & Robinson, S. (1946). Decline in the rates of sweating of men working in severe heat. *Am. J. Physiol.* **147**, 370–8.

Givoni, B. (1963). Estimation of the effect of climate on Man: Development of a new thermal index. Res. Rep. to UNESCO, Israel Institute of Technology, Haifa.

(1964). A new method for evaluating industrial heat exposure and maximum permissible work load. *Int. J. Bioclim. Biometeor.* **8**, 115–24.

Givoni, B. & Belding, H. S. (1962). The cooling efficiency of sweat evaporation. In *Biometeorology II*, ed. Tromp, S. W., pp. 304–14. London: Pergamon.

Givoni, B. & Berner-Nir, E. (1967). *a*, Expected sweat rate as a function of metabolism, environmental factors and clothing; *b*, The thermal effects of solar radiation; *c*, Experimental study of the physiological effects of solar

References

radiation. Rep. to U.S. Dept. Hlth Ed. Welfare Grant BSS-OH-ISR-2, Israel Institute of Technology, Haifa.

Givoni, B. & Rim, Y. (1962). Effect of the thermal environment and psychological factors upon subjects' responses and performance of mental work. *Ergonomics* **1**, 99–114.

Glaser, E. M. & Lee, T. S. (1953). Activity of human sweat glands during exposure to cold. *J. Physiol., Lond.* **122**, 59–65.

Glickman, N., Inouye, T., Keeton, R. W. & Fahnstock, M. K. (1950). Physiologic examination of the Effective Temperature index. *Heat. Pip. Air Condit.* **22**, 157–64.

Glickman, N., Inouye, T., Telser, S. E., Keeton, R. W., Hick, F. K. & Fahnstock, M. K. (1947). Physiological adjustments of human beings to sudden change in environment. *Trans. Am. Soc. Heat. Vent. Engrs* **53**, 327–55.

Goldman, R. F., Green, E. B. & Iampietro, P. F. (1965). Tolerance of hot wet environments by resting men. *J. appl. Physiol.* **20**, 271–7.

Goldsmith, R., Fox, R. H. & Hampton, I. F. G. (1967). Effects of drugs on heat acclimatization by controlled hyperthermia. *J. appl. Physiol.* **22**, 301–4.

Goodman, A. B. & Wolf, A. V. (1969). Insensible water loss from human skin as a function of ambient vapor concentration. *J. appl. Physiol.* **26**, 203–7.

van Graan, C. H. (1969). The determination of body surface area. *S. Afr. med. J.* **43**, 952–9.

van Graan, C. H. & Wyndham, C. H. (1964). Body surface area in human beings. *Nature, Lond.* **204**, 998.

Gregory, H. S. & Rourke, E. (1957). *Hygrometry*. London: Crosby Lockwood.

Guibert, A. & Taylor, C. L. (1952). Radiation area of the human body. *J. appl. Physiol.* **5**, 24–37.

Gurney, R. & Bunnell, I. L. (1942). A study of the reflex mechanism of sweating in the human being; effect of anesthesia and sympathectomy. *J. clin. Invest.* **21**, 269–74.

Haines, G. F. & Hatch, T. (1952). Industrial heat exposures – evaluation and control. *Heat. Vent.* **49**, 93–104.

Hall, J. F. & Klemm, F. K. (1963). Insensible weight loss of clothed resting subjects in comfortable temperatures. *J. appl. Physiol.* **18**, 1188–92.

Halliday, E. C. & Hugo, T. J. (1962). Measurement of the surface area of the human body. *Nature, Lond.* **193**, 584.

(1963). The photodermoplanimeter. *J. appl. Physiol.* **18**, 1285–9.

Hammel, H. T. (1955). Thermal properties of fur. *Am. J. Physiol.* **182**, 369–76.

(1965). Neurones and temperature regulation. In *Physiological Controls and Regulations*, ed. Yamamoto, W. S. & Brobeck, J. R., pp. 71–97. Philadelphia: Saunders.

(1968). Regulation of internal body temperature. *A. Rev. Physiol.* **30**, 641–710.

Hammel, H. T., Jackson, D. C., Stolwijk, J. A. J., Hardy, J. D. & Stromme, S. B. (1963). Temperature regulation by hypothalamic proportional control with adjustable set temperature. *J. appl. Physiol.* **18**, 1146–54.

Hanifan, D. T., Blockley, W. V., Mitchell, M. B. & Strudwick. P. H. (1963). Physiological and psychological effects of overloading fallout shelters. Final report, Contract No. OCD-OS-62-137, Office of Civil Defense, Washington, D.C.

References

Hardy, J. D. (1949). Heat transfer. In *Physiology of Heat Regulation and the Science of Clothing*, ed. Newburgh, L. H., pp. 79–108. Philadelphia: Saunders. Facsimile (1968), New York: Hafner.

(1961). Physiology of temperature regulation. *Physiol. Rev.* **41**, 451–606.

Hardy, J. D. & DuBois, E. F. (1938). The technic of measuring radiation and convection. *J. Nutr.* **15**, 461–75.

Hardy, J. D. & Stolwijk, J. A. J. (1966). Partitional calorimetric studies of Man during thermal transients. *J. appl. Physiol.* **21**, 1799–806.

Harrison, J. & MacKinnon, P. C. B. (1962). The palmar sweat index and the effect of stress and hormonal activity on it. *J. Physiol., Lond.* **162**, 5 *P*.

Harrison, J., MacKinnon, P. C. B. & Monk-Jones, M. E. (1962). Behaviour of the palmar sweat glands before and after operation. *Clin. Sci.* **23**, 371–7.

Hatch, T. F. (1963). Assessment of heat stress. In *Temperature, its Measurement and Control in Science and Industry*, Vol. 3, Pt 3, ed. Hardy, J. D., pp. 307–18. New York: Reinhold.

Heerd, E. & Ohara, K. (1962). Untersuchungen uber die Wasserdampfabgabe kleiner Hautflächen beim Menschen. I. Die Abhängigkeit vom Wasserdampfdruck in der umgebenden Luft bei normaler Hauttemperatur. *Pflügers Arch. ges. Physiol.* **276**, 32–41.

Heerd, E. & Opperman, C. (1966). Seasonal adaptation of insensible water loss in a thermo-neutral environment. In *Human Adaptability and its Methodology*, ed. Yoshimura, H. & Weiner, J. S., pp. 142–47. Tokyo: Jap. Soc. Prom. Sci.

Hellon, R. F. (1963). Peripheral circulation in Man: Local effects of temperature. *Br. med. Bull.* **19**, 141–4.

Hellon, R. F. & Crockford, G. W. (1959). Improvements to the globe thermometer. *J. appl. Physiol.* **14**, 649–50.

Hellon, R. F. & Lind, A. R. (1956). Observations on the activity of sweat glands with special reference to influence of ageing. *J. Physiol., Lond.* **133**, 132–44.

(1958). The influence of age on peripheral vasodilatation in a hot environment. *J. Physiol., Lond.* **141**, 262–72.

Hellon, R. F., Lind, A. R. & Weiner, J. S. (1956). The physiological reactions of men of two age groups to a hot environment. *J. Physiol., Lond.* **133**, 118–31.

Henschel, A., Taylor, H. L. & Keys, A. (1943). Persistance of heat acclimatization in Man. *Am. J. Physiol.* **140**, 321–5.

Hermansen, L. & Andersen, K. L. (1965). Aerobic work capacity in young Norwegian men and women. *J. appl. Physiol.* **20**, 425–31.

Herrington, L. P., Winslow, C.-E. A. & Gagge, A. P. (1937). The relative influence of radiation and convection upon vasomotor temperature regulation. *Am. J. Physiol.* **120**, 133–43.

Hertig, B. A. (1960). Effects of immersion in water on thermal sweating. Thesis for D.Sc., Faculty of the Graduate School of Public Health, University of Pittsburgh.

Hertig, B. A. & Belding, H. S. (1963). Evaluation of health hazards. In *Temperature, its Measurement and Control in Science and Industry*, Vol. 3, Pt 3, ed. Hardy, J. D., pp. 347–55. New York: Reinhold.

Hertig, B. A., Belding, H. S., Kraning, K. K., Batterton, D. L., Smith, C. R. & Sargent, F. (1963). Artificial acclimatization of women to heat. *J. appl. Physiol.* **18**, 383–6.

References

Hertig, B. A., Riedesel, M. L. & Belding, H. S. (1961). Sweating in hot baths. *J. appl. Physiol.* **16**, 647–51.

Hertig, B. A. & Sargent, F. (1963). Acclimatization of women during work in hot environments. *Fedn Proc. Fedn Am. Socs exp. Biol.* **22**, 810–13.

Hertzman, A. B. (1963). Regulation of cutaneous circulation during body heating. In *Temperature, its Measurement and Control in Science and Industry*, Vol. 3, Pt 3, ed. Hardy, J. D., pp. 559–70. New York: Reinhold.

Hertzman, A. B. & Ferguson, I. D. (1959). Failure in temperature regulation during progressive dehydration. Tech. Rep. No. 59–398, Wright Air Development Center, Ohio.

Hey, E. N. (1968). Small globe thermometers. *J. scient. Instrum.* **1**, 955–7 & corrigendum p. 1260.

van Heyningen, R. E. & Weiner, J. S. (1952*a*). A comparison of arm-bag sweat and body sweat. *J. Physiol., Lond.* **116**, 395–403.

(1952*b*). The effect of arterial occlusion on sweat composition. *J. Physiol., Lond.* **116**, 404–13.

Hick, F. K. (1956). See oral discussion, p. 234, Belding & Hatch, 1956.

Hick, F. K., Keeton, R. W. & Glickman, N. (1938). Physiologic response of Man to environmental temperature. *Trans. Am. Soc. Heat. Vent. Engrs* **44**, 145–58.

Hill, A. V. & Howarth, J. V. (1959). The reversal of chemical reactions in contracting muscle during an applied stretch. *Proc. R. Soc.* B **151**, 169–93.

Hill, L. (1921). Cooling and warming of the body by local application of cold and heat. *J. Physiol., Lond.* **54**, cxxxvii–cxxxviii.

Hilpert, R. (1933). Warmeabgabe von geheizten Drahten und Rohren. *Forsch. Geb. Ingenieurwes*, **4**, 215. Cited by Jakob, 1949.

Hodgman, C. D. (1965). *Handbook of Chemistry and Physics*. Cleveland: Chemical Rubber Co.

Höfler, W. (1968). Changes in regional distribution of sweating during acclimatization to heat. *J. appl. Physiol.* **25**, 503–6.

Houberechts, A., Lavenne, F. & Patigny, J. (1958). Le travail humain aux températures élevées. *Maroc méd.* **37**, 328–45.

Houghten, F. C. & Yagloglou, C. P. (1923). Determining lines of equal comfort. *Trans. Am. Soc. Heat. Vent. Engrs* **29**, 163–75.

(1924). Cooling effect on human beings produced by various air velocities. *Trans. Am. Soc. Heat. Vent. Engrs* **30**, 193–209.

Hutchinson, J. C. D. & Brown, G. D. (1969). Penetrance of cattle coats by radiation. *J. appl. Physiol.* **26**, 454–64.

Iampietro, P. F. & Goldman, R. F. (1965). Tolerance of men working in hot humid environments. *J. appl. Physiol.* **20**, 73–6.

Ingersoll, L. R., Zobel, O. J. & Ingersoll, A. C. (1954). *Heat Conduction*. Madison: University of Wisconsin Press.

Issekutz, B., Hetenyi, G. & Diosy, A. (1950). Contributions to the physiology of sweat secretion. *Archs int. Pharmacodyn. Thér.* **83**, 133–42.

Jackson, D. C. & Schmidt-Nielsen, K. (1964). Countercurrent heat exchange in the respiratory passages. *Proc. natn Acad. Sci. U.S.A.* **51**, 1192–7.

Jacquez, J. A., Huss, J., McKeehan, W., Dimitroff, J. M. & Kuppenheim, H. F. (1955*a*). Spectral reflectance of human skin in the region 0.7–2.6 μ. *J. appl. Physiol.* **8**, 297–9.

References

Jacquez, J. A., Kuppenheim, H. F., Dimitroff, J. M., McKeehan, W. & Huss, J. (1955 b). Spectral reflectance of human skin in the region 235–700 μm. *J. appl. Physiol.* **8**, 212–19.

Jakob, M. (1949, 1957). *Heat Transfer*, Vols 1 & 2. New York: Wiley.

Johnson, R. E., Pitts, G. C. & Consolazio, F. C. (1944). Factors influencing chloride concentration in human sweat. *Am. J. Physiol.* **141**, 575–89.

Joyce, J. P., Blaxter, K. L. & Park, C. (1966). The effect of natural outdoor environments on the energy requirements of sheep. *Res. vet. Sci.* **7**, 342–59.

Kaufman, W. C. (1963). Human tolerance limits for some thermal environments of aerospace. *Aerospace Med.* **34**, 889–96.

Kerslake, D. McK. (1955). Factors concerned in the regulation of sweat production in Man. *J. Physiol., Lond.* **127**, 280–96.

— (1962). Heat loss in space. *Proc. 1st Internat. Symp. on Basic Environmental Problems of Man in Space*, ed. Bjurstedt, H., pp. 153–60. Vienna: Springer-Verlag.

— (1963). Errors arising from the use of mean heat exchange coefficients in the calculation of the heat exchanges of a cylindrical body in a transverse wind. In *Temperature, its Measurement and Control in Science and Industry*, Vol. 3, Pt 3, ed. Hardy, J. D., pp. 183–90. New York: Reinhold.

— (1964). A heated manikin for studies on air ventilated clothing. FPRC Memo No. 214, Ministry of Defence, London.

— (1967a). Assessment of the sensible heat transfer properties of conditioned clothing. *Proc. XVII Congress of the International Astronautical Federation*, ed. Malina, F. J. pp. 175–81.

— (1967b). A sensitive recording balance for weighing human subjects. *Med. Engng* **5**, 570–89.

— (1968). Thermal radiation in the investigation of cutaneous vasomotor and sudomotor control. In *Thermal Problems in Aerospace Medicine*, ed. Hardy, J. D. (Agardograph 111). Maidenhead: Technivision.

Kerslake, D. McK. & Brebner, D. F. (1970). Maximum sweating at rest. In *Physiological and Behavioural Temperature Regulation*, ed. Hardy, J. D., Gagge, A. P. & Stolwijk, J. A. J., pp. 139–51. Springfield: Thomas.

Kerslake, D. McK. & Waddell, J. L. (1958). The heat exchanges of wet skin. *J. Physiol., Lond.* **141**, 156–63.

King, G. (1945). Permeability of keratin membranes to water vapour. *Trans. Faraday Soc.* **41**, 479–87.

Koch, W., Jennings, B. H. & Humphreys, C. M. (1960). Sensation responses to temperature and humidity under still-air conditions in the comfort range. *Trans. Am. Soc. Heat. Refrig. Air-Condit. Engrs* **2**, 264–82.

Kraning, K. K., Belding, H. S. & Hertig, B. A. (1966). Use of sweating rate to predict other physiological responses to heat. *J. appl. Physiol.* **21**, 111–17.

Kuno, Y. (1934, 1956). *The Physiology of Human Perspiration*. 1st ed. Springfield: Thomas; 2nd ed. London: Churchill.

Ladell, W. S. S. (1945). Thermal sweating. *Br. med. Bull.* **3**, 175–9.

— (1951). Rapid recovery of sweating in the arm after arterial occlusion. *J. Physiol., Lond.* **115**, 69P.

— (1964). Terrestrial animals in humid heat: Man. In *Handbook of Physiology*, Section 4, ed. Dill, D. B., pp. 625–59. Washington: Am. Physiol. Soc.

References

Larose, P. (1947). The effect of wind on the thermal resistance of clothing with special reference to the protection given by coverall fabrics of various permeabilities. *Can. J. Res.* A **25**, 169–90.

Lee, D. H. K. (1964). Terrestrial animals in dry heat: Man in the desert. In *Handbook of Physiology*, Section 4, ed. Dill, D. B., pp. 551–82. Washington: Am. Physiol. Soc.

Lee, D. H. K. & Henschel, A. (1963). Evaluation of thermal environment in shelters. Rep. No. TR–8, U.S. Dept Hlth Ed. Welfare, Washington, D.C.

Lee, D. H. K. & Henschel, A. (1965). Effects of physiological and clinical factors on response to heat. *Ann. N.Y. Acad. Sci.* **134**, 743–9.

Lee, D. H. K. & Vaughan, J. A. (1964). Temperature equivalent of solar radiation in Man. *Int. J. Bioclim. Biomet.* **8**, 61–9.

Lentz, C. P. & Hart, J. S. (1960). The effect of wind and moisture on heat loss through the fur of newborn caribou. *Can. J. Zool.* **38**, 669–88.

Levine, S. Z. & Marples, E. (1930). Insensible perspiration in infancy and in childhood; basal metabolism and basal insensible perspiration of normal infant; statistical study of reliability and of correlation. *Am. J. Dis. Child.* **40**, 269–84.

Lewis, H. E., Foster, A. R., Mullan, B. J., Cox, R. N. & Clark, R. P. (1969). Aerodynamics of the human microenvironment. *Lancet* **1**, 1273–7.

Lewis, W. K. (1922). *Mech. Engng* **44**, 445. Cited by Jakob, 1949.

Libet, B. (1945). Estimation of thermal insulation of clothing by measuring increases in girth of the wearer. AAF Rep. No. TSEAL-5H-5-241, Air Tech. Serv. Command.

Lichton, I. J. (1957). Osmotic pressure of human sweat. *J. appl. Physiol.* **11**, 422–4.

Lind, A. R. (1962). Individual variations in physiological responses to continuous work in comfortable and hot environments. Rep. No. 62/6, Nat. Coal Board Physiol. Panel, London.

(1963 a). A physiological criterion for setting thermal environmental limits for everyday work. *J. appl. Physiol.* **18**, 51–6.

(1963 b). Physiological effects of continuous or intermittent work in the heat. *J. appl. Physiol.* **18**, 57–60.

(1964). *Heat Stress and Heat Disorders*, Leithead, C. S. & Lind, A. R. London: Churchill.

Lind, A. R. & Hellon, R. F. (1957). Assessment of physiological severity of hot climates. *J. appl. Physiol.* **11**, 35–40.

Lind, A. R., Hellon, R. F., Weiner, J. S. & Jones R. M. (1955). Tolerance of men to work in hot, saturated environments with reference to mines rescue operations. *Br. J. ind. Med.* **12**, 296–303.

Lind, A. R., Leithead, C. S. & McNicol, G. W. (1968). Cardiovascular changes during syncope induced by tilting men in the heat. *J. appl. Physiol.* **25**, 268–76.

Lipkin, M. & Hardy, J. D. (1954). Measurement of some thermal properties of human tissues. *J. appl. Physiol.* **7**, 212–17.

Lloyd, D. P. C. (1959). Secretion and reabsorption in sweat glands. *Proc. natn. Acad. Sci. U.S.A.* **45**, 405–9.

Lobitz, W. C. & Daniels, F. (1961). Skin. *A. Rev. Physiol.* **32**, 207–28.

References

Loewy, A. & Wechselmann, W. (1911). Zur Physiologie und Pathologie des Wasserwechäels und der Wärmeregulation seitens des Hautorgans. *Virchows Arch. path. Anat. Physiol.* **206**, 79–121.

Löfstedt, B. (1961). Beklädnadens inverkan på människans tolerans för höga temperaturer och luftfuktigheter. *Nord. hyg. Tidskr.* **42**, 247. Cited by Lind, 1964.

(1966). *Human Heat Tolerance*. Lund: Berlingska Boktryckeriet.

Lorenzi, R. J., Herrington, L. P. & Winslow, C.-E. A. (1946). The influence of an interior coating of aluminium paper on internal thermal conditions and heat economy. *Heat. Pip. Air Condit.* **18**, 109–14.

McAdams, W. H. (1942). *Heat Transmission*, 2nd ed. New York: McGraw-Hill.

McArdle, B., Dunham, W., Hollong, H. E., Ladell, W. S. S., Scott, J. W., Thomson, M. L. & Weiner, J. S. (1947). The prediction of the physiological effects of warm and hot environments. Rep. No. RNP 47/391, Med. Res. Council, London.

McConnell, W. J. & Houghten, F. C. (1923). Some physiological reactions to high temperatures and humidities. *Trans. Am. Soc. Heat. Vent. Engrs* **29**, 131–63.

McConnell, W. J., Houghten, F. C. & Yagloglou, C. P. (1924). Air motion – high temperatures and various humidities – reactions on human beings. *Trans. Am. Soc. Heat. Vent. Engrs* **30**, 199–224.

McCook, R. D., Wurster, R. D. & Randall, W. C. (1965). Sudomotor and vasomotor responses to changing environmental temperature. *J. appl. Physiol.* **20**, 371–8.

McCutchan, J. W. & Taylor, C. L. (1950). Respiratory heat exchange with varying temperature and humidity of inspired air. A. F. Tech. Rep. No. 6023, Wright-Patterson Air Force Base, Ohio.

MacDonald, D. K. C. & Wyndham, C. H. (1950). Heat transfer in Man. *J. appl. Physiol.* **3**, 342–64.

McGuire, J. H. (1953). *Heat Transfer by Radiation*. DSIR Fire Res. Sp. Rep. No. 2. London: H.M.S.O.

Machle, W. & Hatch, T. F. (1947). Heat: Man's thermal exchanges and physiological responses. *Physiol. Rev.* **27**, 200–27.

McLaughlin, J. T. & Sonnenschein, R. R. (1963). Response of human sweat glands to local heating. *J. invest. Derm.* **41**, 27–9.

McLean, J. A. (1963). The partition of insensible losses of body weight and heat from cattle under various climatic conditions. *J. Physiol., Lond.* **167**, 427–47.

Macpherson, R. K. (1960). *Physiological Responses to Hot Environments*. Med. Res. Council Sp. Rep. No. 298. London: H.M.S.O.

Macpherson, R. K. & Newling, P. S. B. (1954). Salt concentration and rate of evaporation of sweat. *J. Physiol., Lond.* **123**, 74P.

Mali, J. W. (1956). The transport of water through the human epidermis. *J. invest. Derm.* **27**, 451–69.

Minard, D. (1964). Effective temperature scale and its modifications. Res. Rep. MR. 005.01-0001.01 No. 6, N.M.R.I., Bethesda.

Minard, D., Belding, H. S. & Kingston, J. R. (1957). Prevention of heat casualties. *J. Am. med. Ass.* **165**, 1813–18.

303

References

Minard, D., Kingston, J. R. & van Liew, H. D. (1960). Heat stress in working spaces of an aircraft carrier. Res. Rep. MR 005.01-0001.01 No. 3. N.M.R.I. Bethesda.

Minard, D. & O'Brien, R. L. (1964). Heat casualties in the navy and marine corps 1959–1962 with appendices on the field use of the wet bulb-globe temperature index. Res. Rep. MR 005.01-0001.01 No. 7. N.M.R.I., Bethesda.

Missenard, A. (1935). Theorie simplifiee du thermometre resultant – thermostat resultant. *Chauff. Vent.* **12**, 347. Cited by Bedford, 1964.

(1944). Influence des conditions thermiques ambiantes sur la capacité de travail, la morbidité et la mortalité des ouvriers et la construction des lieux de travail. Com. d'Organisation du Batimen et des Travaux Publics. Circular, series L No. 8, Feb. 5th. Cited by Bedford, 1964.

(1948). Equivalence thermique des ambiances, équivalences de passage, équivalences de séjours. *Chal. Ind.* **24**, 159–72, 189–98.

Mitchell, D., Wyndham, C. H., Atkins, A. R., Vermeulen, A. J., Hofmeyr, H. S., Strydom, N. B. & Hodgson, T. (1968). Direct measurements of the thermal responses of nude resting men in dry environment. *Pflügers Arch. ges. Physiol.* **303**, 324–43.

Mitchell, D., Wyndham, C. H. & Hodgson, T. (1967). Emissivity and transmittance of excised human skin in its thermal emission wave band. *J. appl. Physiol.* **23**, 390–4.

Mitchell, D., Wyndham, C. H., Vermeulen, A. J., Hodgson, T., Atkins, A. R. & Hofmeyr, H. S. (1969). Radiant and convective heat transfer of nude men in dry air. *J. appl. Physiol.* **26**, 111–18.

Mole, R. H. (1948). The relative humidity of the skin. *J. Physiol., Lond.* **107**, 399–411.

Molnar, G. W. & Rosenbaum, J. C. (1963). Surface temperature measurement with thermocouples. In *Temperature, its Measurement and Control in Science and Industry*, Vol. 3, Pt 3, ed. Hardy, J. D. pp. 3–11. New York: Reinhold.

Monash, S. (1957). Location of the superficial epithelial barrier to skin penetration. *J. invest. Derm.* **29**, 367–76.

Monash, S. & Blank, H. (1958). Location and reformation of the epithelial barrier to water vapour. *Archs Derm.* **78**, 710–14.

Montagna, W. (1962). *The Structure and Function of Skin*, 2nd ed. New York: Academic Press.

Morimoto, T., Slabochova, Z., Naman, R. K. & Sargent, F. (1967). Sex differences in physiological reactions to thermal stress. *J. appl. Physiol.* **22**, 526–32.

Muller, E. A. (1962). Occupational work capacity, *Ergonomics* **5**, 445–452.

Murlin, J. R. & Burton, A. C. (1935). Human calorimetry. 1. A semi-automatic respiration calorimeter. J. Nutr. 9, 233–60.

Nelson, N., Eichna, L. W., Horvath, S. M., Shelley, W. B. & Hatch, T. F. (1947). Thermal exchanges of Man at high temperatures. *Am. J. Physiol.* **151**, 626–52.

Newburgh, L. H. & Johnston, M. W. (1942). The insensible loss of water. *Physiol. Rev.* **22**, 1–18.

References

Newburgh, L. H., Johnston, M. W., Lashmet, C. H. & Sheldon, J. M. (1937). Further experiences with the measurement of heat production from insensible loss of weight. *J. Nutr.* **13**, 203–21.

Nielsen, B. (1966). Regulation of body temperature and heat dissipation at different levels of energy production in Man. *Acta physiol. scand.* **68**, 215–27.

(1968). Thermoregulatory responses to arm work, leg work and intermittent work. *Acta physiol. scand.* **72**, 25–32.

(1969). Thermoregulation in rest and exercise. *Acta physiol. scand.* Supplement 323.

Nielsen, B. & Nielsen, M. (1962). Body temperature during work at different environmental temperatures. *Acta physiol. scand.* **56**, 120–9.

(1965 a). On the regulation of sweat secretion in exercise. *Acta physiol. scand.* **64**, 314–22.

(1965 b). Influence of passive and active heating on the temperature regulation of Man. *Acta physiol. scand.* **64**, 323–31.

Nielsen, M. (1938). Die Regulation der Korpertemperatur bei Muskelarbeit. *Scand. Arch. Physiol.* **79**, 193–230.

Nishi, Y. & Gagge, A. P. (1970). Moisture permeation of clothing – a factor governing thermal equilibrium and comfort. *Trans. Am. Soc. Heat. Refrig. Air-Condit. Engrs* **76**, pt. 1.

O'Brien, J. P. (1947). A study of miliaria rubra, tropical anhidrosis and anhidrotic asthensia. *Br. J. Derm.* **59**, 125–158.

Ogawa, T. (1970). Local effect of skin temperature on threshold concentration of sudorific agents. *J. appl. Physiol.* **28**, 18–22.

Ohara, K., Kondo, M. & Ogino, J. (1963). Fluctuations in insensible perspiration. *Jap. J. Physiol.* **13**, 441–53.

Peirce, F. T. & Rees, W. H. (1945). The transmission of heat through textile fabrics. *Shirley Inst. Mem.* **19**, 341–64.

Peiss, C. N., Randall, W. C. & Hertzman, A. B. (1956). Hydration of the skin and its effect on sweating and evaporative heat loss. *J. invest. Derm.* **26**, 459–70.

Penman, H. L. (1955). *Humidity.* London: Institute of Physics.

Peter, J. & Wyndham, C. H. (1966). Activity of the human eccrine sweat gland during exercise in a hot humid environment before and after acclimatization. *J. Physiol., Lond.* **187**, 583–94.

Pfliederer, H. & Less, L. (1935). Die Klimatischen Ansprüche an die Atemwege des menschlichen Korpers. *Bioklim. Beibl.* **2**, 1. Cited by McCutchan & Taylor, 1950.

Phelps, E. B. & Vold, A. (1934). Studies in ventilation. I. Skin temperature as related to atmospheric temperature and humidity. *Am. J. publ. Hlth* **24**, 959–70.

Piironen, P. (1970). Sinusoidal signals in the analysis of heat transfer in the body. In *Physiological and Behavioural Temperature Regulation*, ed. Hardy, J. D., Gagge, A. P. & Stolwijk, J. A. J., pp. 358–66. Springfield: Thomas.

Pinson, E. A. (1942). Evaporation from human skin with sweat glands inactivated. *Am. J. Physiol.* **137**, 492–503.

Potter, B. (1966). The physiology of the skin. *A Rev. Physiol.* **28**, 159–76.

References

Provins, K. A., Hellon, R. F., Bell. C. R. & Hiorns, R. W. (1962). Tolerance to heat of subjects engaged in sedentary work. *Ergonomics* **5**, 93–7.

Pugh, L. G. C. E. & Chrenko, F. A. (1966). The effective area of the human body with respect to direct solar radiation. *Ergonomics* **9**, 63–7.

Randall, W. C. (1947). Local sweat gland activity due to direct effects of radiant heat. *Am. J. Physiol.* **150**, 365–71.

Randall, W. C., Deering, R. & Dougherty, I. (1948). Reflex sweating and the inhibition of sweating by prolonged arterial occlusion. *J. appl. Physiol.* **1**, 53–9.

Randall, W. C. & Peiss, C. N. (1957). The relationship between skin hydration and the suppression of sweating. *J. invest. Derm.* **28**, 435–41.

Randall, W. C., Rawson, R. O., McCook, R. D. & Peiss, C. N. (1963). Central and peripheral factors in dynamic thermoregulation. *J. appl. Physiol.* **18**, 61–4.

Randall, W. C., Wurster, R. D., McCook, R. D. & Brockhouse, J. E. (1965). Vascular and sweating responses to regional heating. *Archs envir. Hlth* **11**, 430–41.

Rapp, G. M. (1970). Convective mass transfer and the coefficient of evaporative heat loss from the human skin. In *Physiological and Behavioural Temperature Regulation*, ed. Hardy, J. D., Gagge, A. P. & Stolwijk, J. A. J., pp. 55–80. Springfield: Thomas.

Rawson, R. O. & Hardy, J. D. (1967). Sweat inhibition by cutaneous cooling in normal, sympathectomised and paraplegic Man. *J. appl. Physiol.* **22**, 287–91.

Rawson, R. O. & Randall, W. C. (1961). Vascular and sweating responses to regional heating of the body surface. *J. appl. Physiol.* **16**, 1006–10.

Reader, S. R. & Whyte, H. M. (1951). Tissue temperature gradients. *J. appl. Physiol.* **4**, 396–402.

Richardson, H. B. (1926). Clinical calorimetry. XL. The effect of the absence of sweat glands on the elimination of water from the skin and lungs. *J. biol. Chem.* **67**, 397–411.

Robinson, S. (1944). CMR Interim Rep. No. 12, O.S.R.D. Cited by Hatch, 1963. (1949). Physiological adjustments to heat. In *Physiology of Heat Regulation and the Science of Clothing*, ed. Newburgh, L. H., pp. 193–239, Philadelphia: Saunders. Facsimile (1968), New York: Hafner. (1965). Mechanism of sweating in work. *Archs envir. Hlth* **11**, 454–9.

Robinson, S. & Gerking, S. D. (1947). Thermal balance of men working in severe heat. *Am. J. Physiol.* **149**, 476–88.

Robinson, S., Gerking, S. D., Turrell, E. S. & Kincaid, R. K. (1950a). Effect of skin temperature on salt concentration of sweat. *J. appl. Physiol.* **2**, 654–62.

Robinson, S., Kincaid, R. K. & Rhany, R. K. (1950). Effect of salt deficiency on the salt concentration in sweat. *J. appl. Physiol.* **3**, 55–62.

Robinson, S., Meyer, F. R., Newton, J. L., Ts'Ao, C. H. & Holgersen, L. O. (1965). Relations between sweating, cutaneous blood flow and body temperature in work. *J. appl. Physiol.* **20**, 575–82.

Robinson S. & Robinson, A. H. (1954). Chemical composition of sweat. *Physiol. Rev.* **34**, 202–20.

References

Robinson, S. & Turrell, E. S. (1944). Studies of the physiological effects of solar radiation. Int. Rep. No. 11, Dept of Physiol., Indiana Univ. Med. Sch. Cited by Burton & Edholm, 1955.

Robinson, S., Turrell, E. S., Belding, H. S. & Horvath, S. M. (1943). Rapid acclimatization to work in hot climates. *Am. J. Physiol.* **140**, 168–76.

Robinson, S., Turrell, E. S. & Gerking, S. D. (1945). Physiologically equivalent conditions of air temperature and humidity. *Am. J. Physiol.* **143**, 21–32.

Roe, C. F., Hardy, J. D. & Stolwijk, J. A. J. (1967). Thermoregulatory responses to local warming of skin over the spinal column. *Physiologist* **10**, 292.

Roller, W. L. & Goldman, R. F. (1968). Prediction of heat load on Man. *J. appl. Physiol.* **24**, 717–21.

Rothman, S. & Coon, J. M. (1940). Axon reflex responses to acetylcholine in the skin. *J. invest. Derm.* **3**, 79–97.

Rowley, F. B., Jordan, R. C. & Snyder, W. E. (1947). Comfort reactions of workers during occupancy of air conditioned offices. *Trans. Am. Soc. Heat. Vent. Engrs* **53**, 357–64.

Saito, K. (1930). On local sweating on heated area of skin in Man. *J. orient. Med.* **13**, 53.

Saltin, B., Gagge, A. P. & Stolwijk, J. A. J. (1968). Muscle temperature during submaximal exercise in Man. *J. appl. Physiol.* **25**, 679–88.

 (1970). Body temperatures and sweating during thermal transients caused by exercise. *J. appl. Physiol.* **28**, 318–27.

Saltin, B. & Hermansen, L. (1966). Esophageal, rectal and muscle temperature during exercise. *J. appl. Physiol.* **21**, 1757–62.

Sargent, F. (1962). Depression of sweating in Man: so-called sweat gland fatigue. In *Advances in Biology of Skin*, Vol. 3, ed. Montagna, W., Ellis, R. A. & Silver, A. I., pp. 163–212. London: Pergamon.

Schmidt-Nielsen, K. (1963). Heat conservation in counter-current systems. In *Temperature, its Measurement and Control in Science and Industry*, Vol. 3, Pt 3, ed. Hardy, J. D., pp. 143–6. New York: Reinhold.

Schwartz, I. L., Thaysen, J. H. & Dole, V. P. (1953). Urea excretion in human sweat as a tracer for movement of water within the secreting gland. *J. exp. Med.* **97**, 429–37.

Seeley, L. E. (1940). Study of changes in the temperature and water vapor content of respired air in the nasal cavity. *Trans. Am. Soc. Heat. Vent. Engrs* **46**, 259–89.

Senay, L. C., Christensen, M. & Hertzman, A. B. (1961). Cutaneous vasodilatation elicited by body heating in calf, forearm, cheek and ear. *J. appl. Physiol.* **16**, 655–9.

Sendroy, J. & Cecchini, L. P. (1954). Determination of human body surface area from height and weight. *J. appl. Physiol.* **7**, 1–12.

Sendroy, J. & Collison, H. A. (1960). Nomogram for determination of human body surface area from height and weight . *J. appl. Physiol.* **15**, 958–9.

Sheard, C., Williams, M. M. D. & Horton, B. T. (1939). Skin temperatures of the extremities and effective temperature. *Trans. Am. Soc. Heat. Vent. Engrs* **45**, 153–60.

Sibbons, J. L. H. (1966). Assessment of thermal stress from energy balance considerations. *J. appl. Physiol.* **21**, 1207–17.

References

Sibbons J. L. H. (1970). Coefficients for evaporative heat transfer. In *Physiological and Behavioral Temperature Regulation*, ed. Hardy, J. D., Gagge, A. P. & Stolwijk, J. A. J., pp. 108–38. Springfield: Thomas.

Siple, P. A. & Cochran, M. I. (1944). Conference on principles of environmental stress on soldiers, pp. 14–17. Office of the Quartermaster General, Environmental Protection Section. Cited by Burton & Edholm, 1955.

Smith, F. E. (1952). Effective temperature as an index of physiological stress. Rep. No. 53/728, Med. Res. Council, London.

— (1955). *Indices of Heat Stress*. Med. Res. Council Memo No. 29. London: H.M.S.O.

Snellen, J. W. (1966). Mean body temperature and the control of thermal sweating. *Acta physiol. pharmac. néerl.* **14**, 99–174.

Snellen, J. W., Mitchell, D. & Wyndham, C. H. (1970). Heat of evaporation of sweat. *J. appl. Physiol.* **29**, 40–4.

Spells, K. E. (1960). Theoretical model of the air ventilated suit. The case when the permeable material is of cylindrical form. Rep. No. 155, R.A.F. Inst. Av. Med., Farnborough.

— (1961). Theoretical model of the air ventilated suit. The case when the boundary condition at the outer surface is that of heat flux dependent on a heat transfer coefficient. FPRC Rep. No. 1137, Ministry of Defence, London.

— (1966). The dynamic insulation effect in air ventilated suits. A discussion of previously reported experimental results with correlations to assist in practical problems of design to required standards of performance in protection against heat or cold. FPRC Rep. No. 1266, Ministry of Defence, London.

Spells, K. E. & Blunt, O. J. (1962). The air ventilated suit. Experiments to investigate the improvement by thermal insulation due to air flow through the material. FPRC Rep. No. 1202, Ministry of Defence, London.

— (1965). The air ventilated suit. Further experiments by two different methods to study the gain in thermal insulation due to air flow through the material. FPRC Rep. No. 1233, Ministry of Defence, London.

Stoll, A. & Greene, L. C. (1959). Relationship between pain and tissue damage due to thermal radiation. *J. appl. Physiol.* **14**, 373–82.

Stolwijk, J. A. J. & Hardy, J. D. (1966a). Partitional calorimetric studies of responses of Man to thermal transients. *J. appl. Physiol.* **21**, 967–77.

— (1966b). Temperature regulation in Man – a theoretical study. *Pflügers Arch. ges. Physiol.* **291**, 129–62.

Stolwijk, J. A. J., Hardy, J. D. & Rawson, R. O. (1962). Thermoregulatory responses in Man following sudden changes in environmental temperature. *J. appl. Physiol.* **21**, 967–77.

Stolwijk, J. A. J., Saltin, B. & Gagge, A. P. (1968). Physiological factors associated with sweating during exercise. *Aerospace Med.* **39**, 1101–5.

Sulzberger, M. B., Herrmann, F., Borota, A. & Strauss, M. B. (1953). Studies on sweating. VI. On the urticariogenic properties of human sweat. *J. invest. Derm.* **21**, 293–303.

Sunderman, F. W. (1941). Persons lacking sweat glands; hereditary ectodermal dysplasia of the anhidrotic type. *Archs intern. Med.* **67**, 846–54.

Taylor, C. L. (1954). Preliminary analysis of evaporation from the human body. Thermal biotechnology project, Memo No. 4, U.C.L.A.

Taylor, C. L. & Buettner, K. (1953). Influence of evaporative forces upon skin temperature dependency of human perspiration. *J. appl. Physiol.* **6**, 113–23.

Taylor, P. F. (1956). Middle East Trials: Meteorological observations. CSEE Rep. No. 67, Ministry of Supply, London.

Thauer, R. (1965). Circulatory adjustments to climatic requirements. In *Handbook of Physiology*, Section 2, Vol. 3, ed. Hamilton, W. F., pp. 1921–66. Washington: Am. Physiol. Soc.

Thaysen, J. H. & Schwartz, I. L. (1955). Fatigue of the sweat glands. *J. clin. Invest.* **34**, 1719–25.

Timbal, J., Colin, J., Guieu, J. D. & Boutelier, C. (1969). A mathematical study of thermal losses by sweating in Man. *J. appl. Physiol.* **27**, 726–30.

Tregear, R. T. (1965). Hair density, wind speed and heat loss in mammals. *J. appl. Physiol.* **20**, 796–801.

(1966). *Physical Functions of Skin.* London: Academic Press.

Trolle, C. (1937). A study of the insensible perspiration in Man and its nature. *Skand. Arch. Physiol.* **76**, 225–46.

Underwood, C. R. & Ward, E. J. (1966). The solar radiation area of Man. *Ergonomics* **9**, 155–68.

Veghte, J. H. & Webb, P. (1961). Body cooling and response to heat. *J. appl. Physiol.* **16**, 235–8.

Vernon, H. M. (1932). The measurement of radiant heat in relation to human comfort. *J. ind. Hyg. Toxicol.* **14**, 95–111.

Vernon, H. M. & Warner, C. G. (1932). The influence of the humidity of the air on capacity for work at high temperatures. *J. Hyg., Camb.* **32**, 431–63.

Walker, J. E. C. & Wells, R. E. (1961). Heat and water exchange in the respiratory tract. *Am. J. Med.* **30**, 259–67.

Webb, P. (1951). Air temperatures in respiratory tracts of resting subjects in cold. *J. appl. Physiol.* **4**, 378–82.

(1955). Respiratory heat loss in cold. *Fedn Proc. Fedn Am. Socs exp. Biol.* **14**, 486–7.

(1959). Human thermal tolerance and protective clothing. *Ann. N.Y. Acad. Sci.* **82**, 714–23.

(1963). Pain limited heat exposures. In *Temperature, its Measurement and Control in Science and Industry*, Vol. 3, Pt 3, ed. Hardy, J. D., pp. 245–50, New York: Reinhold.

(1964). *Bioastronautics data book.* Washington: N.A.S.A.

(1968). The space activity suit: an elastic leotard for extravehicular activity. *Aerospace Med.* **39**, 376–83.

Webb, P. & Annis, J. F. (1967). Biothermal responses to varied work programs in men kept thermally neutral by water cooled clothing. Rep. No. CR-739, Washington: N.A.S.A.

(1968). Cooling required to suppress sweating during work. *J. appl. Physiol.* **25**, 489–93.

Webb, P., Garlington, L. N. & Schwarz, M. J. (1957). Insensible weight loss at high skin temperatures. *J. appl. Physiol.* **11**, 41–4.

References

Weiner, J. S. (1945). The regional distribution of sweating. *J. Physiol., Lond.* **104**, 32–40.

Weiner, J. S. & van Heyningen, R. E. (1952). Relation of skin temperature to salt concentration of general body sweat. *J. appl. Physiol.* **4**, 725–33.

Weinman, K. P., Slabochova, Z., Bernauer, E. M., Morimoto, T. & Sargent, F. (1967). Reactions of men and women to repeated exposure to humid heat. *J. appl. Physiol.* **22**, 533–8.

Wiley, F. H. & Newburgh, L. H. (1931). The relationship between the environment and the basal insensible loss of weight. *J. clin. Invest.* **10**, 689–701.

Williams, C. G. & Wyndham, C. H. (1968). The problem of optimum acclimatization. *Int. Z. angew. Physiol.* **26**, 298–308.

Williams, C. G., Wyndham, C. H. & Morrison, J. F. (1967). Rate of loss of acclimatization in summer and winter. *J. appl. Physiol.* **22**, 21–6.

Wilkie, D. R. (1964). Heat, work and chemical change in muscle. *Proc. R. Soc.* B **160**, 476–80.

Winslow, C.-E. A. & Gagge, A. P. (1941). Influence of physical work on physiological reactions to the thermal environment. *Am. J. Physiol.* **134**, 664–81.

Winslow, C.-E. A., Gagge, A. P. & Herrington, L. P. (1940). Heat exchange and regulation in radiant environments above and below air temperature. *Am. J. Physiol.* **131**, 79–82.

Winslow, C.-E. A. & Greenberg, L. (1935). The thermointegrator – a new instrument for the observation of thermal interchanges. *Heat. Pip. Air Condit.* **7**, 41–3.

Winslow, C.-E. A., Herrington, L. P. & Gagge, A. P. (1937a). Physiological reactions of the human body to varying environmental temperatures. *Am. J. Physiol.* **120**, 1–22.

(1937b). Physiological reactions of the human body to various atmospheric humidities. *Am. J. Physiol.* **120**, 288–99.

(1937c). Relations between atmospheric conditions, physiological reactions and sensations of pleasantness. *Am. J. Hyg.* **26**, 103–15.

Woodcock, A. H. (1962). Moisture transfer in textile systems. *Text. Res. J.* **32**, 628–33.

Woodcock, A. H. & Breckenridge, J. R. (1965). A model description of thermal exchange for the nude man in hot environments. *Ergonomics* **8**, 223–35.

Woodcock A. H., Powers, J. T. & Breckenridge, J. R. (1956). Man's thermal balance in warm environments. Tech. Rep. No. EP-30, Office of the Quartermaster General, Natick.

Woodcock, A. H., Pratt, R. L. & Breckenridge, J. R. (1952). Heat exchange in hot environments. Rep. No. EP-183, Office of the Quartermaster General, Natick.

(1960). Theory of the globe thermometer. Res. Study Rep. No. BP-7, Office of the Quartermaster General, Natick.

Wurster, R. D. & McCook, R. D. (1969). Influence of rate of change in skin temperature on sweating. *J. appl. Physiol.* **27**, 237–40.

Wyndham, C. H. (1954). Effect of acclimatization on circulatory responses to high environmental temperatures. *J. appl. Physiol.* **4**, 383–95.

(1962). Tolerable limits of air conditions for men at work in hot mines. *Ergonomics* **5**, 115–22.

(1967). Effect of acclimatization on the sweat rate/rectal temperature relationship. *J. appl. Physiol.* **22**, 27–30.

Wyndham, C. H., Allan, A. McD., Bredell, G. A. G. & Andrew, R. (1967*a*). Assessing the heat stress and establishing the limits for work in a hot mine. *Br. J. ind. Med.* **24**, 255–71.

Wyndham, C. H. & Atkins, A. R. (1968). A physiological scheme and mathematical model of temperature regulation in Man. *Pflügers Arch. ges. Physiol.* **303**, 14–30.

Wyndham, C. H., v.d. Bouwer, W., Devine, M. G. & Paterson, H. E. (1952*a*). Physiological responses of African laborers at various saturated air temperatures, wind velocities and rates of energy expenditure. *J. appl. Physiol.* **5**, 290–8.

Wyndham, C. H., v.d. Bouwer, W., Devine, M. G., Paterson, H. E. & MacDonald, D. K. C. (1952*b*). Examination of use of heat exchange equations for determining changes in body temperature. *J. appl. Physiol.* **5**, 299–307.

Wyndham, C. H., v.d. Bouwer, W., Paterson, H. E. & Devine, M. G. (1953). Examination of heat stress indices: Usefulness of such indices for predicting responses of African mine laborers. *Am. Med. Ass. Archs ind. Hyg.* **7**, 221–33.

Wyndham, C. H. & Jacobs, G. E. (1957). Loss of acclimatization after six days in cool conditions on the surface of a mine. *J. appl. Physiol.* **11**, 197–8.

Wyndham, C. H., Morrison, J. F. & Williams, C. G. (1965). Heat reactions of male and female Caucasians. *J. appl. Physiol.* **20**, 357–64.

Wyndham, C. H., Strydom, N. B., Morrison, J. F., du Toit, F. D. & Kraan, J. G. (1954). Responses of unacclimatized men under stress of heat and work. *J. appl. Physiol.* **6**, 681–6.

Wyndham, C. H., Strydom, N. B., Morrison, J. F., Williams, C. G., Bredell, G. A. G., Maritz, J. S. & Munro, A. (1965*b*). Criteria for physiological limits for work in heat. *J. appl. Physiol.* **20**, 37–45.

Wyndham, C. H., Strydom, N. B., Morrison, J. F., Williams, C. G., Bredell, G. A. G. & Peter, J. (1966). Fatigue of the sweat gland response. *J. appl. Physiol.* **21**, 107–10.

Wyndham, C. H., Strydom, N. B., Williams, C. G. & Heyns, A. (1967*b*). An examination of certain individual factors affecting heat tolerance of mine workers. *Jl S. Afr. Inst. Min. Metall.* **68**, 79–91.

Wyndham, C. H., Williams, C. G., Morrison, J. F., Heyns, A. J. A. & Siebert, J. (1968). Tolerance of very hot humid environments by highly acclimatized Bantu at rest. *Br. J. ind. Med.* **25**, 22–39.

Wyndham, C. H., Williams, C. G., Morrison, J. F. & Heyns, A. J. A. (1967*c*). The tolerance of acclimatized men at rest and at work of very high temperatures and humidities. *Jl S. Afr. Inst. Min. Metall.* **68**, 92–100.

Yaglou, C. P. (1927). Temperature, humidity and air movement in industries: the effective temperature index. *J. ind. Hyg.* **9**, 297–309.

(1947). A method for improving the effective temperature index. *Trans. Am. Soc. Heat. Vent. Engrs* **53**, 307–13.

(1949). Indices of comfort. In *Physiology of Heat Regulation and the Science of Clothing*, ed. Newburgh, L. H., pp. 277–87. Philadelphia: Saunders. Facsimile (1968). New York: Hafner.

References

Yaglou, C. P. (1950). Estimation of radiant heat, equivalent temperature and effective temperature corrected for radiation. (Committee on atmospheric comfort, chairman, Yaglou, C. P.). Am. J. publ. Hlth 40, Year Book, pp. 141–3.

Yaglou, C. P. & Miller, W. E. (1925). Equivalent conditions of temperature, humidity and air movement determined with individuals normally clothed. Trans. Am. Soc. Heat. Vent. Engrs 31, 59–70.

Yaglou, C. P. & Minard, D. (1956). Contr. Rep. to ONR N5-ori-7665. Cited by Minard, 1964.

(1957). Control of heat casualties at military training centers. Am. med. Ass. Archs ind. Hlth 16, 302–16.

Zöllner, G., Thauer, R. & Kaufmann, W. (1955). Der insensible Gewichtsverlust als Funktion der Umweltbedingungen. Die Abhängigkeit der Hautwasserabgabe von der Hauttemperatur bei verschiedenen Temperaturen und Wasserdampfdrucker der umgebender Luft. Pflügers Arch. ges. Physiol. 260, 261–73.

INDEX

Index

Index